COSPAR – IAU – IUTAM

Satellite Dynamics

Symposium São Paulo/Brazil
June 19-21, 1974

Editor G.E.O. Giacaglia

Springer-Verlag
Berlin Heidelberg New York 1975

Editor
Professor G. E. O. GIACAGLIA
Instituto Astronomico e Geofisico, São Paulo/Brazil

Executive Editor
Dr. A. C. STICKLAND
COSPAR Editorial Office, Northwood/England

With 86 Figures

ISBN-13: 978-3-642-46321-1 e-ISBN-13: 978-3-642-46319-8
DOI: 10.1007/978-3-642-46319-8

Softcover reprint of the hardcover 1st edition 1975

Library of Congress Cataloging in Publication Data

Symposium on Satellite Dynamics, 4th, São Paulo, Brazil, 1974.

Satellite dynamics. Bibliography: p. Includes index.

1. Artificial satellites-Dynamics-Congresses.
I. Giacaglia, Giorgio E. O., 1935- ed. II. International Council of Scientific Unions. Committee on Space Research. III. International Astronomical Union.
IV. International Union of Theoretical and Applied Mechanics. V. Title.

TL1050.S95 1974 629.41'1 75-2057

Preface

This volume includes original papers presented at the 4th Symposium on Satellite Dynamics held at the XII Annual Plenary Meeting of COSPAR. At a time where it might be thought that very few problems were left unsolved in celestial mechanics, we discover that new and more challenging questions must be answered. The precision of observations reaches the centimeter level and physical phenomena which had been disregarded come into play. We need a better treatment of atmospheric drag, radiation forces, and a better knowledge of the earth's gravitational field. Time has to be precisely defined as well as reference systems, including improved values for precision and nutation. The question of resonances introduced by nonzonal harmonics was to be carefully investigated. Numerical integration techniques must be optimized and means of controlling their errors improved. Analytical techniques must be made appropriate for computer processing. Presently existing methods of solutions of differential equations of interest to celestial mechanics are getting cumbersome as all these new facts come to light. It is clear that entirely new and more effective methods are necessary. These methods must, among other requirements, take into account the essential nonlinear character of the equations. Finally, the motion about the center of mass of a satellite is becoming an essential need for the thorough understanding and description of the orbital motion.

Several of the above questions are reviewed and discussed in this volume and represent all important additions to the reference literature of scholars and students in celestial mechanics.

We are grateful to COSPAR, IAU and IUTAM for their support and express our deepest appreciation to the São Paulo State Research Foundation (FAPESP) who has provided substantial financial support for travel and maintenance expenses of the invited lecturers.

São Paulo, June 1974

Giorgio E. O. Giacaglia

Participants

<u>Argentine</u>: F. Lopez

<u>Belgium</u>: J. Vercheval

<u>Brazil</u>:
J.B. Andrade, D. Basu, E.W. Benjamin, J.P. Gravier, F. Hatschbach, J.F. Mello, P.F. Mesquita, F. de Mendonca, A. Rios-Neto, C. Patel, C.S. Pereira, J.R. Stremikis, R. Vilhena de Moraes, H.I. Weber

<u>CSSR</u>: L. Sehnal

<u>France</u>:
B. Balmino, X. Berger, N. Capitaine, A. Cazenave, D. Gambis, Z. Niemirowicz, F. Novel

<u>Germany</u>: G.E. Morfill, H. Seeger

<u>Italy</u>: G. Colombo

<u>Japan</u>: G.I. Hori, Y. Kozai

<u>UK</u>: M.R. Richards

<u>USA</u>:
V.R. Bond, J.B. Eades, Jr., A.H. Jazwinski, S. Klosko, J.H. Lieske, C.P. Martin, F. Morrison, P. Nacozy, J. Ryan, J.W. Siry, B. Tapley, W.R. Webber

Contributors

Balmino, G.
G.R.G.S., Observatoire de Meudon, France

Berger, X.
G.R.G.S., Section Astrophysics, Observatoire de Meudon, France.

Culp, R.D.
Department of Aerospace Engineering Sciences, University of Colorado, Boulder, CO 80302, U.S.A.

Gambis, D.
G.R.G.S., Observatoire de Meudon, France

Gaposchkin, E.M.
Center for Astrophysics, Harvard College Observatory and Smithsonian Astrophysical Observatory, Cambridge, MASS. 02138 U.S.A.

Gravier, J.P.
Department of Electrical Engineering, Pontifíca Universidade Católica do Rio de Janeiro, Rio de Janeiro, Brazil

Grigorevsky, V.M.
Moscow

Hipkins, C.
Business and Technological Systems, Inc., Seabrook, MD 20801, U.S.A.

Hori, G.I.
Department of Astronomy, University of Tokyo, Bunkyo-ku, Tokyo, Japan

Jazwinski, A.H.
Business and Technological Systems, Inc., Seabrook, MD 20801, U.S.A.

Kolesnik, S.Ya.
Moscow

Kozai, Y.
Department of Astronomy, University of Tokyo, Bunkyo-ku, Tokyo, Japan

Lohmeier, P.
Institut B fur Mechanik Tech. Universitat Munchen, BRD Germany

Marchal, C.
L'Office d'Etudes et de Recherches Aérospatiales, Chatillon, France

Mulholland, J.D.
Groupe de Recherche de Géodésie Spatiale, Meudon, France, and The University of Texas at Austin, Austin, TX 78712, U.S.A.

Nacozy, P.E.
Department of Aerospace, The University of Texas at Austin, Austin, TX 78712, U.S.A.

Rios-Neto, A.
Escola Politénica, Departmento of Mechanical Engineering, University of São Paulo, 01000 São Paulo, Brazil

Schutz, B.E.
Department of Aerospace Engineering and Engineering Mechanics, The University of Texas at Austin, Austin, TX 78712, U.S.A.

Sehnal, L.
Astronomical Institute of the Czechoslovak Academy of Sciences, Ondrejov, Czechoslovakia

Tapley, B.D.
Department of Aerospace Engineering and Engineering Mechanics, The University of Texas at Austin, Austin, TX 78712, U.S.A.

Velez, C.E.
NASA/Goddard Space Flight Center, Greenbelt, MD, U.S.A.

Vilhena de Moraes, R.
Aeronautics Institute of Technology, Astronomical Observatory, 12200 São José dos Campos, São Paulo, Brazil

Walter, H.G.
Astronomisches Rechen-Institut, Heidelberg, Germany

Weber, H.I.
Departmento de Engenharia Mecânica, Universidade Estadual de Campinas 13100 Campinas, São Paulo, Brazil

Contents

Fourth Session Thursday, June 20, 1974, Afternoon

Chairman: Y. Kozai, Japan

Fifth Session Friday, June 21, 1974, Morning

Chairman: J. Kovalevsky, France

Sixth Session Friday, June 21, 1974, Afternoon

Chairman: E.M. Gaposchkin, U.S.A.

Analytical Theories of the Motion of Artificial Satellites

GEN-ICHIRO HORI and YOSHIHIDE KOZAI
University of Tokyo, Tokyo, Japan

Abstract

The present paper gives a general review of various analytical theories of the motion of artificial satellites presented for these fifteen years or so. Most theories treat the main problem (i.e. problem of the oblateness perturbations), and they are classified according to a) intermediaries adopted, b) perturbation techniques used, and c) orders of approximations carried out. We next review the theories that treat of other sources of perturbations such as tesseral harmonics of the earth's potential, luni-solar attraction, drag of the earth's atmosphere, solar radiation, and so on. All of them are of the second order in magnitude, but are still very important especially in view of secular and long-period perturbations. The resonance cases (i.e. problem of the critical inclination and motion of 24-hours satellites) are also treated by many authors. The orders of perturbations needed for analytical theories to attain an accuracy of 10^{-6} and 10^{-8} are shown in concluding the review.

Motion of Artificial Satellites in Celestial Mechanics. The motion of artificial earth satellites presents very interesting dynamical problems in celestial mechanics. As contrasted with traditional problems of the motion of planets, asteroids, and natural satellites where orbital parameters such as eccentricity e and inclination i are generally small, these parameters in the motion of artificial satellites take any values depending on their initial conditions so that the expansion in powers of them is generally not valid. In addition, a very short time scale of satellites' motion (say, 2 hours for one revolution) yields rapid motions both in the nodal and the apsidal lines of the orbit (say, several months for a circulation of them).

Hence long-period perturbations in e and i have similar periods of several months, then the handling of the long-period perturbations differs considerably from that of planetary theories, and is rather similar to the lunar theory: these perturbations are to be represented in trigonometrical series in time rather than power series as usually used in planetary theories where "long-period" merges in "secular" in practical sense. In other words, long-period terms are to be treated as proper "periodic" perturbations.

On the other hand the force function U of the equation of motion of an earth satellite is rather easy to handle, making a big contrast to the planetary case. In fact, the main part of U is as simple as

$$U = \frac{\mu}{r} - \frac{\mu J_2 R^2}{r} P_2(\sin \beta) \quad ,$$

$$(\mu = \text{grav. const.} \times M_{\oplus}) \tag{1}$$

where R is the equatorial radius of the earth and β the latitude of the satellite referred to the equatorial plane, $J_2 \sim 10^{-3}$ is a constant associated with the dynamical oblateness of the earth. Moreover, since there is only one fast variable (the mean anomaly of the satellite), we have no difficulty associated with commensurability which is a feature in the planetary case. This situation together with the ease of handling U makes the motion of earth satellites rather accessible and many papers have been published on this subject during the past fifteen years.

Main Problem. Intermediaries.

The motion of earth satellites with a force function slightly different from Eq. (1) and of the form

$$U = \frac{\mu}{r} \; \frac{\mu J_2 R^2}{r^3} P_2 \,(\sin \beta) \; - \; \frac{\mu J_3 R^3}{r^4} P_3 \,(\sin \beta)$$

$$- \; \frac{\mu J_4 R^4}{r^5} P_4 \,(\sin \beta) \qquad (2)$$

is called the main problem. In Eq. (2), J_3, $J_4 \sim 10^{-6}$ are constants associated with the dynamical figure of the earth. The force function of the main problem consists of the leading zonal harmonics of the geopotential. The additional attractions coming from the other zonal and nonzonal harmonics of the geopotential as well as attractions due to the sun and the moon, drag of the earth's atmosphere, solar radiation pressure, etc. can be taken into account each separately after the main problem is solved, because each of these additional perturbations are as small as 10^{-6} and their coupling effects of the magnitude 10^{-12} can be neglected. Although the same is true for the motion with the force function (1), the main problem has many additional features of dynamical interest due to the added terms factored by J_3 and J_4 in U. Some of these features will be discussed later.

As mentioned above, the rapid motions of the nodal and apsidal lines of the orbit of an earth satellite suggest the use of an intermediate orbit or an intermediary that have nearly such sidereal, draconic, and anomalistic periods as actually produced by the force function (1). The force function U_0 of an intermediary is such that a) it permits the exact solution of the equations of motion, b) the orbit produced by U_0 is as close as possible to the real orbit, and c) the orbit produced by U_0 is as simple as possible. Clearly the items b) and c) are not compatible with each other.

According to the item a) the form

$$U_0 = f(r) + \frac{g(\beta)}{r^2} \tag{3}$$

is suggested where f and g are arbitrary. Then the item c) restricts the functional form of f and g, leading to

$$U_0 = \frac{\mu}{r} - J_2\left[c_1(\sin^2\beta - c_2)\frac{R^2}{r^2} + c_3\frac{R}{r} + c_4\frac{R^3}{r^3}\right] . \tag{4}$$

simply because hyper-elliptic integrals are otherwise required in the solution. Here c's are parameters independent of r and β. By a proper choice of a set of these parameters, an intermediary associated with U_0 can incorporate the main part of the motions of nodal and apsidal lines. However, the choice is not unique, and the more important is the manner in which the c's are treated in the analysis (11): they may be treated as absolute constants (10,23), or as depending on the orbital parameters (1). This distinction of handling the parameters c's has considerable effect in the development of the perturbation theories based on these intermediaries, and thus has great theoretical significance. A typical intermediary is that of Aksnes with the force function

$$U_0 = \frac{\mu}{r} - \frac{\mu J_2}{p}\frac{R^2}{r^2}P_2(\sin\beta) \tag{5}$$

where p stands for G^2/μ, G being the angular momentum of the motion. This intermediary carries the first order secular perturbations and can be interpreted simply as a slowly rotating Kepler ellipse. It is noted that the disturbing function associated with this intermediary is

$$R = \frac{\mu J_2 R^2}{r^2}P_2(\sin\beta)\left(\frac{1}{r} - \frac{1}{p}\right) + O(J_3, J_4) \quad , \tag{6}$$

being free from the c's as expected from the manner of handling the parameters.

As far as the item a) concerns the most general type of force functions are those which permit the integrability condition of the equations of motion in spheroidal coordinates (25). Along this line of approach, Vinti (24) found a force function of the form

$$U_0 = \frac{\mu}{r} - \frac{J_2 R^2}{r^3} P_2(\sin\beta) - \frac{J_3 R^3}{r^4} P_3(\sin\beta)$$

$$- \frac{J_4^* R^4}{r^5} P_4(\sin\beta) + \sum_{n=3}^{\infty} \frac{(-J_2)^n R^{2n}}{r^{2n+1}} P_{2n}(\sin\beta) \quad ,$$

$$(J_4^* = J_2^2 + J_3^2/J_2) \quad . \tag{7}$$

The intermediary with the force function Eq. (7) is the most accurate one ever presented, and in addition can be expressed in standard elliptic functions. Equations (2) and (7) show that $U-U_0$ is of the second order in magnitude and, in fact, is about $(1-3)J_4$. It is noted that Vinti's force function is free from parameters c's as contrasted w:'.h the force function Eq. (4).

Main Problem. Perturbations.
The use of an accurate intermediary reduces the size perturbations considerably, but it does not necessarily reduce the amount of work involved to obtain those perturbations because of the complexity of expressing the disturbing function by the variables introduced in the intermediary. The item c) required for intermediaries may be justified in view of this situation.

Although the intermediary of Aksnes is not as accurate as that of Vinti, the simpler form of the former makes it possible to establish a perturbation theory through the second order (1) in which, periodic perturbations are obtained through the second order and secular perturbations through the third order. Other second order theories (4*,9,16) and the third order theory by Deprit and Rom (7*) are all based on a very simple intermediary, that is, a Kepler orbit with the force function $U_0 = \mu/r$. Although the Kepler orbit has fixed nodal and apsidal lines and thus seems ineffective to reproduce their fast rotations in the actual motion, its simplicity is highly recognized in carrying out perturbations through higher orders.

In carrying out perturbations based on any intermediary including a Kepler orbit, we rely on a theory of perturbations. As mentioned earlier, long period perturbations are to be properly treated as periodic rather than secular motions in the case of an earth satellite. Hence any theory of perturbations is to allow such a treatment. The von Zeipel method introduced by Brouwer (5) in his first order solution of the main problem is a typical theory of perturbations. The von Zeipel method features: a) the solution of the equations of motion by canonical transformations, and b) the averaging principle to determine suitable transformations. We note that this latter feature makes the proper treatment of long period perturbations possible. However, some disadvantages of the von Zeipel method have been recognized and its revision has been suggested (6,13). In place of implicit canonical transformations with mixed variables (old and new variables mixed), explicit transformations with the

*In these theories the perturbations are not given in closed form with respect to e.

help of Lie series are used in the new approaches. The main part of the development of perturbations is reduced to the evaluation of Poisson brackets, and this simple procedure makes it more accessible to use computers for the literal development of perturbations. The third order theory mentioned above is such an example and a Lie series method is used there.

As mentioned previously, in the theory of the motion of an earth satellite it is desirable to avoid expansion in e. This is possible with the use of f, the true anomaly of the satellite, as the integration variable in the first and second order theories, and possibly in the third order theory with the use of a special function of e defined by an infinite series. But the solution in closed form in e is not expected in the higher order approximations. In this respect, the line of approach of using f as the independent variable proposed by Scheifele and Stiefel (21) is very promising.

When the axial symmetry in the force function of the main problem is taken into account explicitly, the main problem is reduced to a planar motion in r, β plane with the force function

$$U'(r,\beta) = U(r,\beta) - \frac{H^2}{2r^2\cos^2\beta} \qquad (8)$$

where H stands for the component of angular momentum along the axis of symmetry, and U(r,β) the force function Eq. (2). The energy integral of this planar motion reads

$$\frac{1}{2}\left\{p_r^2 + \frac{p_\beta^2}{r^2}\right\} - U(r,\beta) + \frac{H^2}{2r^2\cos^2\beta} = \text{const} = E \quad . \qquad (9)$$

This reduction of the main problem is of theoretical interest. For instance, zero-velocity curve

$$U(r,\beta) = -\frac{H^2}{2r^2\cos^2\beta} = E \tag{10}$$

shows the region of possible motion of the satellite specified by two dynamical parameters H and E.

Main Problem. Small e and i, Small G, Hyperbolic Motion.

As contrasted with the apparent difficulty of large e in the higher order theory of the main problem, there seems to exist no difficulty for small e or i. In fact the position and the velocity of the satellite are well determined when the osculating e or i vanishes. Nevertheless, many papers have been published on the subject of small e or i (13,14,16,17,22). This is mainly because the Delaunay's Type variables, which are adopted to calculate perturbations in the position, and the velocity become singular at vanishing e or i. In fact, some of the unusual behaviors of ℓ (mean anomaly) and g (argument of perigee) in $e \to 0$, and those of g and h (longitude of node) in $i \to 0$ have been discussed (13,17). The former is associated with J_2 and J_4 terms, while the latter with J_3 term in the force function of the main problem. Obviously these singular behaviors of ℓ, g and h cannot be discussed in terms of these variables themselves, but other sets of variables such as Poincaré's are used. However, it is noted that a singular set such as Delaunay's can be safely used in the calculation of $\vec{r}$ and $d\vec{r}/dt$ as recognized in the Lie series method (13).

The case of a large e and a small a (semi-major axis), that is the case of a small angular momentum G seems to offer another difficulty. In fact if G is small, the motion of the satellite approaches rectilinear and the perturbations would increase because of a close approach to the earth (if the earth be an oblate point mass). In view of a finite size of the earth, this case is realized by the motion of a missile rather than a satellite (13,19). This is a case where the averaging principle is not applied. Another case is realized in a hyperbolic motion. The first order theory is discussed by Hori (13).

Main Problem. Critical Inclination.

After eliminating short period perturbations, the main problem is reduced to the motion in G, g variables with a force function of the form

$$U^{*} = J_2 A(G) + J_3 B(G)\cos g + J_4' C(G)\cos 2g \qquad (11)$$

where A, B, C are certain functions of G, and J_4' is a linear form in J_4 and J_2^2.

To find the reduced motion near such a value of G that leads to $dA/dG = 0$ is called the problem of the critical inclination because such a value of G corresponds to a special value of i (64°3). Although the solution can be reduced to the evaluation by quadratures the explicit representation of the solution in terms of the time is desirable, and many papers have been presented on this subject (3,10,12-14,16,18). The amplification of $\sqrt{J_2}$ times of the perturbations of $\vec{r}$ and $d\vec{r}/dt$ near the critical inclination calls for special attention for the ephemeris of the satellite.

Perturbations Due to the Additional Forces

The additional attractions due to the higher zonal harmonics and nonzonal harmonics of the geopotential, the sun and the moon, etc. can be treated as perturbations on the main problem. If the method of canonical transformations are used with Delaunay variables, for instance, the equations of motion for perturbed main problem are

$$\frac{dL_j}{dt} = \frac{\partial F}{\partial \ell_j} \quad , \quad \frac{d\ell_j}{dt} = - \frac{\partial F}{\partial L_j} \quad ,$$

$$F = U + \varepsilon R \quad , \quad \varepsilon = 0(J_2^2) \quad (j = 1,2,3) \qquad (12)$$

where U stands for the force function (2) of the main problem, and εR the disturbing function by any one of the additional forces. It is understood that $\partial R/\partial \ell_j$ and $\partial R/\partial L_j$ are replaced by X_j and Y_j if the force is not conservative. After two canonical transformations, L_j, $\ell_j \rightarrow L_j''$, ℓ_j'', the equations of motion (Eq. 12) are reduced to

$$\frac{dL_j''}{dt} = \frac{\partial F^{**}}{\partial \ell_j''} \quad , \quad \frac{d\ell_j''}{dt} = - \frac{\partial F^{**}}{\partial L_j''} \quad ,$$

$$F^{**} = U^{**}(L_j'') + \varepsilon R^{**}(L_j'', \ell_j'', t) \qquad (13)$$

where εR^{**} is the disturbing function expressed in terms of the new variables. The solution of Eq. (12) via that of Eq. (13) automatically includes the perturbations of the orders $J_2\varepsilon$ as well as that of ε. The terms of the orders $J_3\varepsilon$ and $J_4\varepsilon$ can be omitted. A few theories followed exactly this procedure in the treatment of the air drag perturbations.

An alternative procedure is to solve Eq. (12) directly by considering $U-\mu/r+\varepsilon R$ as the disturbing function, and most theories follow this line of approach.

Perturbations due to nonzonal harmonics as well as higher zonal harmonics have been discussed by many authors (15, etc.). Nonzonal harmonics produce the resonance in the stationary satellites which offers a special topic, but we point out the importance of the luni-solar attractions on high altitude satellites such as the stationary satellites, that is, the theory of the stationary satellites is to incorporate both main nonzonal and luni-solar perturbations. It is noted that nonzonal harmonics of higher orders can produce resonant perturbations of period 10 days say, even for close satellites.

Luni-solar perturbations have been also discussed by many authors (8, in which is furnished good references). The problem resembles the lunar theory, but is easier to handle because of the presence of the predominant J_2 terms. But the situation changes considerably in the stationary satellites where the luni-solar attractions predominate over J_2 attractions. We note also that the orbit of the moon cannot be assumed as an ellipse for any satellite if four figure computations are required in the perturbations. As the luni-solar perturbations contain many long period terms for almost all the satellites, there appear resonant terms of very long periods. The same is true for the perturbations due to the solar radiation pressure.

The solar radiation pressure and the air-drag feature a nontidal and a nonconservative force respectively, and both depend on the area-to-mass ratio of the satellite. For nonspherical satellites as most of the present satellites are, their effective area-to-mass ratios, which depend on the attitude of the satellites, are not known. In addition many parameters which are not exactly known enter into the disturbing forces, so that it is very difficult to compute the perturbations due to these forces by analytical theories. Theoretical interest in the perturbation theory of the radiation pressure is in the treatment of the shadow effects. Although the radiation pressure is a conservative force, it produces a secular change in a, just like a nonconservative force, due to the shadow effects. Another interest may be in Poynting-Robertson effects due to the finite travelling time of the radiation from the sun to the satellites.

The size of disturbing forces of various sources and the perturbations produced are summarized in Table 1 where n is the mean motion of the satellite in degrees per day, A/M the area-to-mass ratio in cgs unit, and k_2 (=0.3) is the Love's number. The factors in the second column represent also the orders of magnitude for short-period perturbations, while the third column shows those for long-period and secular perturbations after 1000 revolutions of satellites.

In order to compute the positions and velocities of satellites with an accuracy of 10^{-6} (10m and 1 cm/sec) it is necessary to evaluate the first order short-period perturbations due to J_2 and J_{22} in the geopotential and those due to the luni-solar terms for the stationary satellite as shown in the fourth column. As for the secular and the long-period perturbations, the first order theory is sufficient for the perturbations caused by nonzonal and higher zonal harmonics, and also by luni-solar, tidal, solar radiation, and air-drag forces. The second order theory is required, however, for the perturbations due to J_2, and luni-

Table 1.

Disturbing force	Factor	lp & s	Orders of Perturb. Needed 10^{-6} accur. sp	10^{-6} accur. lp&s	10^{-8} accur. sp	10^{-8} accur. lp&s
J_2	10^{-3}	1	1	2	2	3
$J_n (n>2)$	10^{-6}	10^{-3}	no	1	1	$1, J_2 J_n$
Nonzonal	$10^{-5} (J_{22}), 10^{-6}$ (others)	10^{-4}	1	1	1	$1 \times J_2, 2$
Luni-solar	$3.1/n^2 \sim 10^{-7}$ (close sat.)	10^{-4}	no	1	1	$1, 1 \times J_2$
	$\sim 10^{-5}$ (station. sat.)	10^{-2}	1	2	1	3,...
Tides	$3.1 \left(\frac{R}{a}\right)^5 k_2/n^2 \sim 10^{-8}$	10^{-5}	no	1	no	1
Solar radiation	$4.6 \times 10^{-8} \left(\frac{a}{R}\right)^2 \frac{A}{M} \sim 10^{-8}$	10^{-2} (mean anomaly)	no	2	no	3
		10^{-5} (other elements)	no	1	no	1
Air-drag	$(A/M)\zeta \sim 10^{-8}$	10^{-2} (mean anomaly)	no	2	no	3
		10^{-5} (other elements)	no	1	no	1

sp = short-period perturbations

lp&s = long-period and secular perturbations

The third column shows orders of magnitude for the long-period and secular perturbations after 1000 revolutions.

solar attractions (for the stationary satellite), and those due to the solar radiation and the air-drag (in the computation of the mean anomaly). These are shown in the fifth column of the table. In addition to the disturbing forces given in the table, there are radiation effects from the surface of the earth, and also the effects of the motion of the earth's equator (precession and nutation).

When an accuracy of 10^{-8} (10cm and 0.1 mm/sec) is needed, analytical theories become much more complicated as shown in the last two columns of the table. Even the computations of short-period perturbations require all the disturbing terms larger than 10^{-8}, and the formulations of secular and long-period perturbations require all terms larger than 10^{-11} in the transformed disturbing function which is derived by eliminating short-period terms. Numerical integration of these transformed equations of motion can be adopted. Here the step of integration is as long as one day even for close satellites since short-period terms have been eliminated. This semi-analytical procedure can adequately handle satellite data with the accuracy of 10^{-8}.

References

1 Aksnes, K.: Astrophys. Norvegica X, n. 4 (1965) 69; Dissertation, Yale University, 1969.

2 Allan, R.R.: Celes. Mech. 2 (1970) 121.

3 Aoki, S.: Astron. J. 68 (1963) 355.

4 Berger, X.: Bulletin G.R.G.S. n. 5 (1972); Bulletin G.R.G.S. n. 8 (1973).

5 Brouwer, D.: Astron. J. 64 (1959) 378.

6 Deprit, A.: Celes. Mech. 1 (1969) 12.

7 Deprit, A.; Rom, A.: Celes. Mech. 2 (1970) 166.

8 Giacaglia, G.E.O.: Celes. Mech. 9 (1974) 239.

9 Gaposhkin, E.: not published (1970).

10 Garfinkel, B.: Astron. J. 63 (1958) 88; Astron. J. 65 (1960) 624; Celes. Mech. 7 (1973) 205.

11 Garfinkel, B.; Aksnes, K.: Astron. J. 75 (1970) 85.

12 Hagihara, Y.: S.A.O. n. 5, 5 (1961) 39.

13 Hori, G.: Astron. J. 65 (1960) 291; Astron. J. 66 (1961) 258; Publ. Astron. Soc. Japan 18 (1966) 287; An article in Recent Advances in Dynamical Astronomy, ed. by B.D. Tapley and V. Szebehely. Dordrecht-Holland: D. Reidel Pub. Co. 1973, 231-249.

14 Iszak, I.: S.A.O. Special Report 90 (1962); Astron. J. 68 (1963) 559.

15 Kaula, W.M.: Theory of Satellite Geodesy. Waltham, U.S.: Blaisdell Pub. Co. 1960.

16 Kozai, Y.: Astron. J. 66 (1961) 132; S.A.O. n. 5, 5 (1961) 53; Astron. J. 67 (1962) 446.

17 Lyddane, R.H.: Astron. J. 68 (1963) 555.

18 Message, P.J., et al.: The Observatory 168 (1962).

19 Peters, C.F.: An article in Periodic Orbits, Stability and Resonances, ed. by G.E.O. Giacaglia. Dordrecht-Holland: D. Reidel Pub. Co. 1970, 469-473.

20 Poritsky, H.: Astron. J. 67 (1962) 212.

21 Scheifele, G.; Stiefel, E.: Canonical Satellite Theory. Report to ESRO. 1972.

22 Smith, O.K.: Astron. J. 66 (1963) 359.

23 Stern, T.E.: Astron. J. 63 (1958) 28.

24 Vinti, J.P.: J. Res. NBS n. 2, 63B (1959) 105; J. Res. NBS n. 1, 70B (1966) 1.

25 Whittaker, E.T.: A Treatise on the Analytical Dynamics of Particles and Rigid Bodies, New York: Dover Pub. 1944, 152.

Time Elements *

PAUL E. NACOZY
University of Texas at Austin U.S.A.

Abstract

Time elements are introduced for use with Sundman time transformations of the type $dt = r^{\alpha}ds$ for satellite equations of motion. Two time elements are presented here, one providing maximum accuracy for $\alpha = 1$, the other for $\alpha = 2$. Time elements and time transformations reduce local truncation error and Liapunov (in track) instability, and provide analytical step size control. Numerical results show accuracy improvements of more than one order of magnitude when time elements are employed with time transformations in the numerical integration of the satellite equations, compared with using time transformations alone.

Time Transformations

Time transformations of the Sundman type (Ref. 1) have the form

$$dt = cr^{\alpha}ds \qquad (1)$$

where t is the time, c is any function of the semi-major axis and eccentricity, α is a positive constant, r the magnitude of the radius-vector, and s the new independent variable.

It is well known that use of the time transformations of Eq. (1) improves rates of convergence of "analytical" series solutions of gravitational systems (Refs. 2 and 3). Also, when equation (1) is used in conjunction with a coordinate transformation, singularities due to collisions may be eliminated from the equations of motion (see, for example, Ref. 4). Several recent

---

*This research was supported by NASA Goddard Grant NGR 44-012-269.

studies (Refs. 5-10) show that the use of the time transformations given by Eq. (1) (without coordinate transformations), with $1 \leq \alpha \leq 2$, substantially improves the accuracy and efficiency of numerical integration of satellite operations of motion. Moreover, it has been shown by many of these same studies that, for a large class of satellite initial conditions, α near or equal to 1.5 provides the maximum accuracy and efficiency (Refs. 7-9).

Time Elements

Time transformations appear to lessen or weaken the dynamical in-track (Liapunov) instability associated with satellite motion as well as provide reduction in local truncation error (Ref. 7) and analytical step size control (Refs. 5,6,8-11). With $\alpha = 2$, Eq. (1) reduces the Keplerian problem to a Liapunov-stable linear oscillator in terms of the independent variable (the true anomaly, if c is properly chosen). However, the reduction of instability in the coordinates is rendered at the expense of the introduction of a differential equation to obtain the time (Eq. 1). The differential equation for the time exhibits some or all, depending on the value of α, of the dynamical instability that was removed from the differential equations for the coordinates.

If the perturbations are small relative to the two-body forces, then some of the instability of the time differential equation may be removed by use of time elements. With no perturbations, Eq. (1) may be integrated in closed form for $\alpha = 1, 1.5, 2$. These integrals define "time elements" (as defined in Ref. 11).

For perturbed motion, the integrals of Eq. (1) are then differentiated using methods of variation of parameters, generating differential equations for the time elements.

The resulting differential equations for the time elements are then integrated by numerical integration. A significant portion of the instability is removed from the equation that defines the time transformation (Eq. 1) by taking into account the two-body solution.

It has been shown by Stiefel and Scheifele (Ref. 11) that a time element may be introduced for the KS equations of motion in the four dimensional u-space. The related time transformation has $\alpha = 1$ and $c = 1$ in Eq. (1).

Often in practice, one may wish to remain in the three dimensional physical space and use only the time transformation of Eq. (1). Hence, two time elements are presented here in the three dimensional physical space. The time elements are valid for all α but each provides maximum accuracy and efficiency for a particular α, one for $\alpha = 1$, and the other for $\alpha = 2$.

The Optimum Time Element for $\alpha = 1$

Kepler's equation may be written in the form

$$t = \tau - \sqrt{\frac{a^3}{\mu}}\; e \sin E \tag{2}$$

where a, e and E are the semi-major axis, eccentricity and eccentric anomaly, respectively, μ the gravitational constant, and where τ is the time element. Equation (2) is then differentiated with respect to s, where s is defined by Eq. (1) and perturbed motion is assumed. The result is

$$\frac{dt}{ds} = \frac{d\tau}{ds} - \sqrt{\frac{a^3}{\mu}}\ e \cos E \frac{dE}{ds} - \sqrt{\frac{a^3}{\mu}} \sin E \frac{de}{ds}$$

$$- \frac{3}{2}\ e \sin E \sqrt{\frac{a}{\mu}} \frac{da}{ds} \quad . \qquad (3)$$

The quantity $\frac{dt}{ds}$ is obtained from Eq. (1) and $\frac{da}{ds} = \frac{da}{dt}\frac{dt}{ds}$ and $\frac{de}{ds} = \frac{de}{dt}\frac{dt}{ds}$ are obtained from the usual Gaussian form of the differential equations for the orbital elements and from Eq. (1). The quantity $\frac{dE}{ds}$ is obtained by considering Kepler's equation as an implicit function of t, t_0, a, e, and E, where t_0 is the time of pericenter passage. Then $\frac{dE}{ds}$ is obtained straightforwardly by use of the methods of differentiation of implicit functions and by use of the Gaussian equations. The results for $\frac{dE}{ds}$ is, after some algebra,

$$\frac{dE}{ds} = \sqrt{\frac{\mu}{a}}\ c\ r^{\alpha-1} + \frac{\sin E}{1 - e \cos E} \frac{de}{ds}$$

$$+ \mu\ c\ ar^{\alpha-1}\ \{(1 - e \cos E)\nabla_v E$$

$$- \sin E\ \nabla_v e\} \cdot \vec{R} \qquad (4)$$

where $\vec{R}$ is the perturbing force vector, and $\nabla_v E$ and $\nabla_v e$ are gradients of the eccentric anomaly and eccentricity with respect to the velocity and are given by

$$\nabla_v E = \frac{1}{\sqrt{\mu^5 a}\ e} \left(\mu^2 \cos E\ \vec{r} - \sqrt{\mu^3 a} \sin E\right.$$

$$\left.\times \{2r + ae \cos E\}\ \dot{\vec{r}}\right)$$

$$\nabla_v e = \frac{1}{\sqrt{\mu a}} \sin E\ \vec{r} + \left\{\frac{2r}{\mu} \cos E - \frac{ae}{\mu} \sin^2 E\right\} \dot{\vec{r}} \qquad (5)$$

where $\vec{r}$ and $\dot{\vec{r}}$ are the position and velocity vectors, respectively.

The Gaussian equations for $\frac{da}{ds}$, $\frac{de}{ds}$ and $\frac{dt_0}{ds}$ may be obtained by use of Poisson's method (see for example, Ref. 12). The results are

$$\frac{da}{ds} = c\, r^{\alpha} \frac{da}{dt} = c\, r^{\alpha} \frac{2a^2}{\mu} \vec{v}\cdot\vec{R}$$

$$\frac{de}{ds} = c\, r^{\alpha} \frac{de}{dt} = c\, r^{\alpha} \left(\frac{1}{\mu a e}\right) \{a^2(1-e^2)\vec{v}\cdot\vec{R} - C\cdot\dot{C}\}$$

$$\frac{dt_0}{dt} = -\mu^{1/2}\cdot \frac{3}{2} \cdot a^{1/2}\ (E - e \sin E) \frac{da}{dt}$$

$$-\mu^{1/2} a^{3/2} \{(1 - e \cos e)\nabla_v E - \sin E\ \nabla_v e\}\cdot\vec{R}$$

where $\vec{v}$ is the velocity vector and C is the magnitude of the angular momentum.

The substitution of Eqs. (1) and (4) into Eq. (3), after rearranging terms, yields the differential equation for the time element τ:

$$\frac{d\tau}{ds} = car^{\alpha-1} + \sqrt{\frac{a^5}{\mu}}\, r^{-1} \sin E \frac{de}{ds}$$

$$+ \frac{3}{2}\, e\sqrt{\frac{a}{\mu}} \sin E \frac{da}{ds}$$

$$+ \sqrt{\frac{a^5}{\mu}}\, ec\, r^{\alpha-1} \cos E \quad .$$

$$\times \{(1 - e \cos E)\nabla_v E - \sin E\ \nabla_v e\}\cdot\vec{R} \quad . \qquad (6)$$

All the terms on the right-hand side of Eq. (6) are of the order of magnitude of the perturbations except the first term. Consequently, if the perturbations are small, τ will be affected mostly by errors in the computation of the first term. To reduce these errors, the semi-major axis in the first term is computed from the total energy of the perturbed system. Let h_k be the unperturbed (Keplerian) energy, h the total energy, and V the perturbing potential. Then

$$a = \frac{-\mu}{2h_k} = \frac{-\mu}{2(h-v)} \quad . \tag{7}$$

For non-conservative systems, the semi-major axis (or the total energy) should be obtained by numerical integration of the Gaussian equation for a (or for the energy). The use of Eq. (7) to determine a is essential for the reduction of Liapunov instability. This is based on the premise that in-track error may be reduced by increasing the accuracy of the computation of the frequency associated with the in-track angular variable (longitude).

For α not equal to one, the first term on the right is also affected by errors in the computation of r. The larger $|\alpha-1|$ becomes the larger is the error in the numerical integration, as will be seen below. For $\alpha = 1$, r disappears, and maximum accuracy is achieved.

When $\alpha = 1$ and $c = \sqrt{\frac{a}{\mu}}$ in Eq. (1), s is the eccentric anomaly for unperturbed (Keplerian) motion and $\tau = \sqrt{\frac{a^3}{\mu}}\, s + t_0 = \sqrt{\frac{a^3}{\mu}}\, E + t_0$.

The Optimum Time Element for $\alpha = 2$

Kepler's equation may be written in the form

$$t = \tau + \sqrt{\frac{a^3}{\mu}}\,(E-f) - \sqrt{\frac{a^3}{\mu}}\, e \sin E \tag{8}$$

where τ is the time element and f is the true anomaly. Differentiating Eq. (8) with respect to s yields

$$\frac{dt}{ds} = \frac{d\tau}{ds} + \frac{3}{2}(E-f)\sqrt{\frac{a}{\mu}}\frac{da}{ds} + \sqrt{\frac{a^3}{\mu}}\left(\frac{dE}{ds} - \frac{df}{ds}\right)$$

$$- \frac{3}{2} e \sin E \sqrt{\frac{a}{\mu}}\frac{da}{ds} - \sqrt{\frac{a^3}{\mu}} \sin E \frac{de}{ds}$$

$$- \sqrt{\frac{a^3}{\mu}} e \cos E \frac{dE}{ds} \quad . \tag{9}$$

The quantity $\frac{df}{ds}$ may be obtained by writing f as an implicit function of a, e, and E, differentiating the function with respect to s, using the Gaussian form of the differential equations for a and e, and using Eq. (4) for $\frac{dE}{ds}$. The result for $\frac{df}{ds}$ is

$$\frac{df}{ds} = c\sqrt{\mu a(1-e^2)} \cdot r^{\alpha-2} + \left\{\frac{a \sin E}{r\sqrt{1-e^2}}\right.$$

$$\left. + \frac{a^2}{r^2}\sqrt{1-e^2} \sin E\right\} \frac{de}{ds} + \sqrt{1-e^2}\, c\, a^2 r^{\alpha-1}$$

$$\times \{(1-e \cos E)\nabla_V E - \sin E\, \nabla_V e\}\cdot\vec{R} \quad . \tag{10}$$

The substitution of Eq. (10) into (9), using Eqs. (1) and (4), and after rearranging terms, yields the differential equation for the time element τ defined by Eq. (8):

$$\frac{d\tau}{ds} = a^2\sqrt{1-e^2}\, c\, r^{\alpha-2} + \frac{1}{r}\sqrt{\frac{a^5}{\mu}}\left\{\frac{\sin E}{\sqrt{1-e^2}}\right.$$

$$\left. + \frac{a}{r}\sin E\sqrt{1-e^2}\right\}\frac{de}{ds} - \frac{3}{2}\sqrt{\frac{a}{\mu}}(E-f-e \sin E)\frac{da}{ds}$$

$$+ c\, r^{\alpha-2}\sqrt{\frac{a^3}{\mu}}\{(1-e \cos E)\nabla_V E - \sin E\, \nabla_V e\}\cdot\vec{R}$$

$$\times \left[a^2\sqrt{(1-e^2)} - r^2\right] \quad . \tag{11}$$

No secular nor mixed secular terms appear in Eq. (11) since the quantity E - f is purely periodic.

All the terms on the right-hand side of Eq. (11) are of the order of the magnitude of the perturbations except the first term. Again, care must be taken in the computation of this term. We have found that the semimajor axis should be computed using Eq. (7) or according to the discussion following Eq. (7). In addition, numerical experiments indicate that the eccentricity in the first-term may be computed directly from the perturbed state vector without significantly affecting the error in τ.

For α not equal to two, the first term on the right is affected by errors in the computation of r. The larger $|\alpha-2|$ becomes, the larger is the error, as will be seen below. For $\alpha = 2$, r disappears and maximum accuracy is achieved.

When $\alpha = 2$ and

$$c = \left[\mu a(1-e^2)\right]^{-1/2}$$

in Eq. (1), s is the true anomaly for unperturbed (Keplerian) motion and

$$\tau = \sqrt{\frac{a^3}{\mu}}\, s + t_0 = \sqrt{\frac{a^3}{\mu}}\, f + t_0 \quad .$$

Numerical Results

To determine the benefit of the time elements, various solutions were obtained by numerical integration using the time transformation of Eq. (1). The value of the time was obtained in three ways: 1) by direct numerical integration of Eq. (1); 2) by direct numerical integration of Eq. (6) using Eq. (2); and, 3) by direct numerical integration of Eq. (11) using Eq. (8). The time transformation was applied to a close earth satellite with eccentricity of 0.3 and included the zonal harmonic perturbations on the satellite. The solutions were obtained over one revolution. Selected solutions were obtained for 30 revolutions with no significant change in the one revolution results that are presented here. The numerical integration used for the integration was a fourth order Runge-Kutta with fixed step size.

The results are summarized in Table 1. For $\alpha = 1$, Eqs. (2) and (6) yield about 1-1/2 orders of magnitude more accuracy than direct integration of Eq. (1) (40 times more accurate). And Eqs. (8) and (11) yield about twice the accuracy of integration of Eq. (1).

For $\alpha = 1.5$, both set (2) and (6) and set (8) and (11) yield about four times the accuracy of integration of Eq. (1).

For $\alpha = 2.0$, Eqs. (8) and (11) yield more than one order of magnitude more accuracy than direct integration of Eq. (1) (20 times more accurate). And Eqs. (2) and (6) yield about twice the accuracy of Eq. (1).

Table 1. Relative accuracy improvements of Eqs. (2) & (6) versus Eq. (1) and Eqs. (8) & (11) versus Eq. (1).

Value of α	Relative accuracy improvement of Eqs. (2) & (6)	Relative accuracy improvement of Eqs. (8) & (11)
$\alpha = 1$	40	2
$\alpha = 1.5$	4	4
$\alpha = 2$	2	20

Conclusions and Continuing Studies

We are currently deriving a time element that provides optimum accuracy for $\alpha = 1.5$ (i.e., more than one order of magnitude improvement compared with using Eq. (1)). It is given in terms of incomplete elliptic integrals and Jacobian elliptic functions and is analogous to Eqs. (8) and (11).

In addition, we are presently simplifying Eqs. (2), (6), (8) and (11) by rewriting them in terms of the state vector, the perturbing forces and the semi-major axis and eccentricity.

Using these time elements for satellite solutions, the time defined by Eq. (1) is now obtainable much more accurately than by direct numerical integration of Eq. (1), at least for the range $1 \leq \alpha \leq 2$ and for gravitational systems with small perturbations relative to the two-body force.

References

1 Sundman, K.F.: Acta. Math. 36 (1912) 105.

2 Hansen, P.A.: Suppl. Comptes Rend. 1 (1856) 121.

3 Gylden, H.: Ofversigt Kongl. Vet. - Akad. Forhandl. 31 n. 1 (1874) 13.

4 Szebehely, V.: Theory of Orbits. New York: Academic Press 1967.

5 Tapley, B.D.; Szebehely, V.; Lewallen, J.M.: AIAA 7 (1969) 1010.

6 Czuchry, A.J.; Pitkin, E.T.: Journal of Spacecraft and Rockets 7 (1970) 882.

7 Nacozy, P.: Bull. Am. Astron. Soc. 5 n. 1 (1973) (Abstract).

8 Walker, D.; Downs, R.; Rusick, A.; Prislin, R.: Trace Documentation. The Aerospace Corporation, 1972 Los Angeles, California, Report No. TR-0059(9320)-1.

9 Wagner, W.F.; Velez, C.E.: GTDS Documentation, NASA Goddard Space Flight Center, 1972, X-552-72-244.

10 Wong, P.: Some Special Perturbation Methods of Integration of Artificial Satellite Orbits, Ph.D. Dissertation, U.C.L.A. (1972).

11 Stiefel, E.L.; Scheifele, G.: Linear and Regular Celestial Mechanics. New York: Springer-Verlag 1971.

12 Giacaglia, G.E.O.: U.S. Naval Weapons Laboratory Technical Memorandum, 1963. No. K-88/63.

An Estimation Procedure for Orbit Determination, Using the K.S. Transformation

ATAIR RIOS-NETO
Escola Politēnica
University of São Paulo, Brazil

Abstract

Initially the orbit estimation problem is formulated. Assuming a perturbed two body physical model, the equations of motion in the mathematical model are treated by using the Kustaanheimo-Stiefel transformation in order to get an analytical solution in parametric space; this solution is explicit in parametric time and the only numerical approximation needed in the propagation of the state is the one to relate parametric to physical time. The observation relationships, at each step of estimation, are expressed in terms of variables of the parametric state so that the problem of determining the orbit is transferred to parametric space. Once the problem of orbit determination is formulated in terms of parametric variables, a procedure is given, making use of the sequential extended Kalman filter. It is expected that the procedure proposed will provide a gain in accuracy as well as a saving in computer work, since no numerical integration of the state is involved.

Introduction

Using a first order Gauss-Markov approximation process to study the results obtained in orbit determination of either a near Earth satellite (2) or a Lunar satellite (4) perturbing forces were obtained which were not well known, and this has motivated the procedure now proposed. Here, the perturbed two body motion of the satellite is transferred to parametric space by use of the Kustaanheimo-Stiefel (K.S.) transformation (1) and the perturbation to the Keplerian motion approximated by a first order Gauss-Markov process. The orbit estimation is then carried out in parametric space by making use of a sequential extended Kalman filter (3). In this first part of the study, only the analytical development of the procedure is presented. It is expected that the procedure now proposed will provide an effective way of determining the orbit of a satellite around a central body for which the potential is not well known. A gain in accuracy is also expected as well as a saving in computer work, since the only numerical approximation is for a single scalar equation and no numerical integration of the whole state is necessary.

Problem Formulation

The orbit determination problem is such that at instants $t_1, t_2, \ldots, t_m$ one is given the observation vectors

$$Y_k = G_k(X_k, t_k) + \varepsilon_k \quad , \quad k = 1, 2, \ldots, m \tag{1}$$

where the subscript k indicates evaluation at time t_k, Y_k being a $(p\times 1)$ observation vector, X_k the state of the satellite, i.e. a $(n\times 1)$ vector of position and velocity components and ε_k a $(p\times 1)$ observation noise vector with known statistical properties, which are taken, as is usually done, as

$$E(\varepsilon_k) = 0 \quad , \quad E(\varepsilon_k \varepsilon_j^T) = R_k \delta_{kj} \tag{2}$$

where, in Eq. (2), R_k is a $(p\times p)$ known diagonal matrix and δ_{kj} a Kroenecker symbol.

Besides the above one knows that the orbit evolutes in physical space according to the following differential equations,

$$\dot{X} = f_0(X,t) + f(X,t) \tag{3}$$

where $f_0(X,t)$ is the portion corresponding to the Kepler motion and $f(X,t)$ is its perturbation, which is not well known. Instead of trying to treat the problem in physical space, one transfers it to parametric space by the use of the K.S. transformation, given below (1),

$$\begin{aligned}
x_1 &= u_1^2 - u_2^2 - u_3^2 + u_4^2 \\
x_2 &= 2(u_1u_2 - u_3u_4) \\
x_3 &= 2(u_1u_3 + u_2u_4) \\
x_4 &= (2/r)(u_1u_5 - u_2u_6 - u_3u_7 + u_4u_8) \\
x_5 &= (2/r)(u_2u_5 + u_1u_6 - u_4u_7 - u_3u_8) \\
x_6 &= (2/r)(u_3u_5 + u_1u_7 + u_4u_6 + u_2u_8) \qquad (4)
\end{aligned}$$

$$dt = rds \qquad (5)$$

where x_i are the components of X such that $\dot{x}_j = x_{j+3}$ for $j = 1,2,3$; u_i are the components of the parametric state vector u, such that $u_j' = u_{j+4}$ for $j = 1,2,3,4$, the prime indicating differentiation with respect to the parametric time s; finally r in Eqs. (4), (5) is the magnitude of the position vector and is such that (1)

$$r = \sum_{j=1}^{4} u_j^2 \quad .$$

To obtain the initial conditions in parametric space, one also needs (1)

$$\begin{aligned}
u_1^2 + u_4^2 &= (1/2)(r + x_1) \\
u_2 &= (x_2u_1 + u_3u_4)/(r + x_1) \\
u_3 &= (x_3u_1 - x_2u_4)/(r + x_1) \qquad (6)
\end{aligned}$$

or

$$\begin{aligned}
u_2^2 + u_3^2 &= (r - x_1)/2 \\
u_1 &= (x_2u_2 + x_3u_3)/(r - x_1) \\
u_4 &= (x_3u_2 - x_2u_3)/(r - x_1) \qquad (7)
\end{aligned}$$

where one takes Eqs. (6), (7) if $x_1>0$ $(x_1<0)$ and chooses u_4 (u_3) arbitrarily at initial time. It is also necessary to compute

$$u_5 = (u_1x_4 + u_2x_5 + u_3x_6)/2$$
$$u_6 = (-u_2x_4 + u_1x_5 + u_4x_6)/2$$
$$u_7 = (-u_3x_4 - u_4x_5 + u_1x_6)/2$$
$$u_8 = (u_4x_4 - u_3x_5 + u_2x_6)/2 \qquad (8)$$

at initial time.

The Problem in Parametric Space

If one applies the K.S. transformation to perturbed two-body motion, there results for the equations of motion (1),

$$u_j'' + \left((M/2r_k) - v_k^2/4\right)u_j = F_j \qquad (9)$$

where $j = 1,2,3,4$, r_k and v_k are magnitudes of position and velocity at time t_k, M is the product of central mass by the gravitational constant and $F_j = F_j(U,s)$ is the perturbing force. Besides Eq. (9) one also has to consider Eq. (5), relating parametric time to physical time, i.e.

$$t' = r \quad . \qquad (10)$$

Since F_j, in Eq. (9), is not well known, one takes the approximation of substituting it by an adaptive first order Gauss-Markov process, i.e. F_j is approximated by e_j, so that

$$e_j = b_j\ e_j + n_j \qquad (11)$$

where Eq. (11) is for $j = 1,2,3,4$ and n_j is Gaussian, not time dependent, independent of the observation noise and of the initial state, $U(t_k) = U_k$. It is assumed that $n^T = (n_1:n_2:n_3:n_4)$ is described by the following statistics,

$$E(n) = 0 \quad , \quad E(n\ n^T) = Q \quad . \qquad (12)$$

From Eq. (11) one gets

$$e_j = \left(\exp(b_j(s-s_k))\right)(e_{jk} + n_j/b_j) - n_j/b_j \qquad (13)$$

where e_{jk} is at time t_k and $j = 1,2,3,4$.

From Eq. (9) there results

$$u_j'' + (M/(2r_k) - v_k/4)u_j = e_j \qquad (14)$$

and the previous equation has a closed form solution which gives, if one associates to time t_k the parametric time $s_k = 0$ (see Eq. (10)),

$$u_j = u_j(U_k, e_{jk}, b_j, n_j, s) \qquad (15)$$

where* U_k is at time t_k and s is such that it corresponds to $t \geq t_k$. But if one now goes back to Eq. (10) with Eq. (15) one gets an equation which can be integrated so that

$$t = t(U_k, e_k, b, n, t_k, s) \qquad (16)$$

where $e_k^T = (e_{1k}:e_{2k}:e_{3k}:e_{4k})$, $b^T = (b_1:b_2:b_3:b_4)$.

If, with the previous results at hand, one goes back to Eqs. (1), those observation relationships would be expressed, in terms of parametric variables, as follows:

$$Y_{k+1} = g_{k+1}(U_k, e_k.b.n.t_k.t_{k+1}, s_{k+1}) + \varepsilon_{k+1} \qquad (17)$$

where $g_{k+1}(.)$ comes from $G_{k+1}(.)$ by substituting X_{k+1} in terms of Eq. (4) and Eq. (15). This completes the formulation of the problem in parametric space.

The Estimation Algorithm

The estimation can be done by using an extended Kalman filter (3,4), where, in a typical interval (t_k, t_{k+1}), the following extended vector is considered for such estimation,

*Notice that $U_k = \left(u_1(t_k):u_2(t_k):\ldots u_8(t_k)\right)$

$$Z^T = (U^T : e^T : b^T) \quad . \tag{18}$$

It is assumed that at time t_k one has previous information on $\bar{Z}_k$ and $\bar{P}_k$, i.e. the nominal state and the state covariance matrix, which is such that

$$\bar{P}_k = E\left[(Z_k-\bar{Z}_k)(Z_k-\bar{Z}_k)/Y_1,Y_2,\ldots,Y_k\right] \quad ; \tag{19}$$

with this information one is able to make the estimation by taking the steps below.

a) At time $t = t_{k+1}$: one takes Eq. (17) and makes a Taylor series expansion about the nominal state, considering only the terms up to first order,

$$Y_{k+1} = g^*_{k+1} + \left(\partial g_{k+1}/\partial Z_k\right)^*(Z_k-\bar{Z}_k)$$

$$+ \left(\partial g_{k+1}/\partial s_{k+1}\right)^*(s_{k+1}-\bar{s}_{k+1})$$

$$+ \varepsilon_{k+1} + \text{h.o.t.} \tag{20}$$

where the asterisk indicates evaluation at the nominal state, i.e. $Z_k = \bar{Z}_k$, $n = 0$, $s_{k+1} = \bar{s}_{k+1}$. If one now goes back to Eq. (16), evaluated at time t_{k+1}, and makes an expansion about the nominal state, one gets

$$t_{k+1} = t^*_{k+1} + \left(\partial t/\partial Z_k\right)^*(Z_k-\bar{Z}_k)$$

$$+ \left(\partial t/\partial s_{k+1}\right)^*(s_{k+1}-\bar{s}_{k+1}) + \text{h.o.t.} \quad . \tag{21}$$

Combining Eqs. (20) and (21), we find

$$y_{k+1} = h_{k+1}\cdot(Z_k-\bar{Z}_k) + \varepsilon_{k+1} \tag{22}$$

where terms of high order have been neglected, h_{k+1} is evaluated at the nominal trajectory and $\bar{s}_{k+1}$ can be determined either by numerical integration of Eq. (5) or by applying an iterative procedure to Eq. (16), where both equations must be considered for nominal values. One should notice that this is the only numerical step needed for the propagation of the state. One can now apply the filter (3),

$$\hat{z}_k = \bar{z}_k + K_{k+1} \cdot y_{k+1} \tag{23}$$

$$P_k = \left(I - K_{k+1} h_{k+1}\right) \bar{P}_k \tag{24}$$

$$K_{k+1} = \bar{P}_k h_{k+1}^T \left(h_{k+1} \bar{P}_k + R_{k+1}\right)^{-1} \tag{25}$$

where, in Eq. (24), I is the identity matrix and R_{k+1} has already been defined in Eq. (2).

b) Propagation of state: if one takes $\bar{z}_k^+ = \hat{z}_k$ and goes back to Eq. (16) with this value and n = 0, one gets $\bar{s}_{k+1}^+$. With these values in Eq. (13) one finally gets $\bar{z}_{k+1}$. To propagate the state covariance matrix, one linearizes Eqs. (13), (15), those corresponding to the fixed coefficients $(b(t) = b(t_k))$ and Eq. (16) about the rectified trajectory, i.e. $\bar{z}_k^+$, s_{k+1}^+, n = 0; one eliminates $(s_{k+1} - \bar{s}_{k+1}^+)$ from the linearized equation that resulted from Eq. (16), so as to get

$$z_{k+1} - \bar{z}_{k+1} = \Gamma_1^*(t_k, t_{k+1}) \cdot (z_k - \bar{z}_k^+) + \Gamma_2^*(t_k, t_{k+1}) \cdot n + \text{h.o.t.} \tag{26}$$

where $\Gamma_1^*(t_k, t_{k+1})$, $\Gamma_2^*(t_k, t_{k+1})$ are only dependent on the nominal values. If one now recalls that n is independent of the initial state, one finds

$$\bar{P}_{k+1} = \Gamma_1^*(t_k, t_{k+1}) P_k \Gamma_1^{*T}(t_k, t_{k+1}) + \Gamma_2^*(t_k, t_{k+1}) Q \Gamma_2^{*T}(t_k, t_{k+1}) \quad . \tag{27}$$

c) Sequence of the estimator: in order to close the loop, one has just to take k = k+1 and go back to (a).

d) Results in physical space: at the end of the processing one has just to make use of Eq. (4) to get the state vector. For the state covariance matrix, one has to linearize Eq. (4) about the nominal trajectory, resulting in

$$(X-\bar{X}) = [L]^{*} \cdot (U-\bar{U}) + \text{h.o.t.} \qquad (28)$$

and

$$E[(X-\bar{X})(X-\bar{X})^{T}] = [L]^{*} P [L]^{*T} \qquad (29)$$

where, in the previous equations, $[L]^{*}$ is the matrix of the partial derivatives that resulted from the linearization, which is, of course, evaluated at the nominal value.

References

1 Stiefel, E.L.; Scheifele, G.: Linear and Regular Celestial Mechanics. New York, Heidelberg, Berlin: Springer-Verlag 1971.

2 Rios-Neto, A.: The estimation of the state and the unmodeled atmospheric drag for near earth satellites. Ph.D. Dissertation, The University of Texas at Austin, March 1973.

3 Jazwinski, A.H.: Stochastic Processes and Filtering Theory. Mathematics in Science and Engineering. Academic Press, v. 64, 1970.

4 Tapley, B.D.: Statistical Orbit Determination Theory. Proceedings of the NATO Advanced Study Institute in Dynamical Astronomy, cortina D'Amprezzo, Italy: Reidel Press, August 1972.

Importance of the Moon-Earth Coupling Effect in the Motion of an Artificial Satellite: Semi-Analytical Computed Solution of This and Similar Problems

X. BERGER
G.R.G.S.
Meudon Observatory, France

Abstract

For a very accurate representation the analytical solution of the motion of an artificial satellite under the simultaneous gravity influence of Earth and Moon one needs to consider coupling the effects not only by secular and long period terms, but also by short period terms. This has been computed and pointed out. We have also computed (and checked by analytical-numerical comparisons) a very effective semi-analytical solution of this problem and similar ones, where the force is small, where its effect is difficult to express analytically (and still much more difficult to mix with the other effects), and where the analytical development is so complicated as to lose its descriptive and operational qualities.

With an analytically computed theory to third order for the Earth gravity field, giving the motion of any Earth artificial satellite with an accuracy of 30 cm, we have computed the development (1) for the influences of Moon and Sun. The force function is written in the form given by Challe and Laclaverie (2). This development has been considered to: the power 2 of a/a^* for solar influence, because we find some terms with an amplitude of 100 m for the position of the perigee (Table 1); the 5th multiple of mean anomaly, because there also exist some large terms with such an amplitude. 11000 terms are so calculated.

As a first step, the moon motion was given by Connaissance des Temps for the mean motion, with some periodic terms to take into account variation and evection. The development of the problem with these periodic terms for Moon motion has been computed.

Coupling by secular terms is fully accomplished between Earth, Moon and Sun, but this is insufficient (Fig. 1).

Coupling by long periodic terms on the angular elements is fully accomplished at all times. That has been made possible by the fact that these elements are only present in sinusoidal functions of time. Coupling by long periodic terms in the metric elements (mainly the long periodic terms in e) is fully accomplished at time = 0, but degrades slowly with time. This requires one to proceed semi-analytically in steps of about six days. To perfect the solution, we consider terms in t^2 to take into account the modification of the secular terms with time.

But with a computed solution of this nature the accuracy in the position of the satellite is imperfect for the lunar influence. Some meters of difference from numerical integrations remain (Figs. 2 and 3).

The first improvement was to obtain a better lunar motion. Thus, we consider the exact motion of the Moon, given by the well known developments of the longitude, latitude and parallax, and we transform them into the orbital system of the satellite in the form:

$$a = a_0 + a_1 \cos(\alpha_1 t + \phi) \qquad \Omega = \Omega_0 + \Omega_1 t$$

$$e = e_0 + e_1 \cos(\varepsilon_1 t + \psi) \qquad \omega = \omega_0 + \omega_1 t$$

$$i = i_0 \qquad M = M_0 + M_1 t$$

The coefficients are modified at each step (Fig. 4). With such a motion, which is better than the one first considered, and sufficiently good to analyse correctly the lunar influence, the accuracy in the position of the satellite is better, but still not perfect.

Then, we have developed and demonstrated that there also exist coupling effects with short periodic terms. We have found that it is possible to find important effects in the observed residuals from e and ω. The problem becomes very difficult to develop analytically. For a small effect, such as lunar influence, the developments are so complicated that each term does not mean anything in itself. It is better to use other means to consider lunar influence and its coupling with other effects.

For this, we have constructed, computed and checked by comparison with numerical integration, a new program for producing orbits. Its place is between the analytical and numerical, the exact mixture being at the choice of the user. Its principle lies in the fact that, for small effects, like the Moon, Sun, air drag and some others, one may calculate numerically the effect on the satellite at any time, not only with the exact position of the satellite, but also with an approximate position.

We start with mean elements and calculate an approximate position of the satellite for an arbitrary given potential.

Few analytical terms are needed to get a position with an accuracy of half a kilometer, although a great number of terms are needed to get a position with an accuracy of 30 cm. So it is very easy to obtain an approximate position. This position is used to calculate numerically the influence of the external forces, that is the forces not used for the potential considered so far. We obtain the effect of these external forces in the form da/dt, de/dt ..., and then, using the Burlisch Stoer method, we extrapolate the mean elements with these derivatives, as well as the secular effects of the initial potential.

Because we extrapolate mean elements and because the external forces are low, we can have a large step (8 days is an optimum step). If the real position of the satellite is not needed at any one step, we proceed to the next step. If we wish then, we can consider only the whole development of the effect of the initial potential (Fig. 5). The speed is about 15 times faster than numerical integration by the Burlisch-Stoer method, and 30 times faster than numerical integration by the Cowell method. This method is good for constructing very long arcs, but not short ones.

If the approximate position is only constructed with only long periodic terms, and then also with short periodic terms, the difference between the two final orbits gives the coupling effect by short periodic terms between Moon and Earth (Fig. 6): it is not negligible.

We have shown the difficulties in analytical development of the influence of the Moon and tried to carry it as far as possible. In such a problem, the computation and comparisons with numerical integration are necessary to demonstrate the correct development which must be used: here, they have indicated the importance of some terms and of some coupling effects which have been neglected by some authors (3-7). For such a difficult problem, though the proposed computed solution is almost correct, the Remora method is perhaps the very effective solution.

References

1 Berger, X.; Boudon, Y.: Théorie analytique programmée de l'influence gravitationnelle de la Lune et du Soleil dans le mouvement des satellites artificiels. Bulletin G.R.G.S. n. 5 (1972).

2 Challe, A.; Laclaverie, J.J.: Fonction perturbatrice et représentation analytique du mouvement d'un satellite. As. and Astro. (1969) 3-15.

3 Kaula, W.: Development of the lunar and solar disturbing functions for a close satellite. Astr. Journal n. 5, 67 (1962).

4 Giacaglia, G.E.O.: Lunar perturbations on artificial satellites on the Earth. S.A.O. Spec. Rep. n. 352 (1973).

5 Kozai, Y.: Lunisolar perturbations with short periods. S.A.O. Spec. Rep. n. 235 (1966).

6 Estes, R.H.: On the analytic lunar and solar perturbations of a near Earth satellite. G.S.F.C. X-645-72-297 (1972).

7 Kozai, Y.: A new method to compute lunisolar perturbations in satellite motions. S.A.O. Spec. Rep. n. 349 (1973).

Table 1. Lunar influence on D1C satellite. Amplitudes of the terms. The smallest terms have been put equal to 0.

Secular terms (rad/sec) Ω: -0.365×10^{-10} ω: 0.465×10^{-10} M: -0.430×10^{-10}

Periodic terms

s	n	p	q	h	k	a	e	i	Ω	ω	ω+M	day	hr	min	
(M	a/a*M*		Ω−Ω*	ω	ω*)										
-1	2	0	0	-2	0	-0.18E-04	0.14E-07	-0.30E-08	0.46E-08	0.16E-06	0.0	0	1	49	
-1	2	0	1	0	0	-0.18E-04	-0.14E-07	-0.19E-08	0.74E-09	0.17E-06	0.0	0	1	50	
-1	2	2	1	0	2	0.19E-04	0.15E-07	0.20E-08	-0.77E.09	-0.17E-06	0.0	0	1	49	
-1	2	2	2	0	2	-0.16E-04	-0.12E-07	-0.33E-08	0.40E-08	0.14E-06	0.0	0	1	49	
0	2	-3	0	0	-2	0.0	0.0	0.0	-0.20E-06	0.26E-06	0.0	8	9	24	
0	2	-2	0	0	-2	0.0	0.0	0.0	-0.15E-05	0.19E-05	0.0	13	6	5	
0	2	-1	0	0	0	0.0	0.0	0.0	-0.10E-05	0.13E-05	0.0	22	21	0	
0	2	1	0	0	0	0.0	0.0	0.0	0.10E-05	-0.13E-05	0.0	22	21	0	
0	2	2	0	0	2	0.0	0.0	0.0	0.15E-05	-0.19E-05	0.0	13	6	5	
0	2	3	0	0	2	0.0	0.0	0.0	0.20E-06	-0.26E-06	0.0	8	9	24	
0	2	-2	0	2	-2	0.0	-0.58E-07	0.0	0.0	0.68E-06	0.0	9	11	42	
0	2	-1	0	2	0	0.0	0.0	0.0	0.0	0.38E-06	0.0	13	13	48	
0	2	0	0	2	0	0.0	-0.88E-06	0.0	0.0	0.10E-04	0.0	33	9	21	
0	2	1	0	2	0	0.0	0.0	0.0	0.0	-0.20E-05	0.0	72	15	29	
0	2	2	0	2	2	0.0	0.13E-06	0.0	0.0	-0.16E-05	0.0	21	23	26	
0	2	-2	0	-2	-2	0.0	0.13E-06	0.0	0.0	0.16E-05	0.0	21	23	26	
0	2	-1	0	-2	0	0.0	0.0	0.0	0.0	0.20E-05	0.0	72	15	29	
0	2	0	0	-2	0	0.0	-0.88E-06	0.0	0.0	-0.10E-04	0.0	33	9	21	
0	2	1	0	-2	0	0.0	0.0	0.0	0.0	-0.38E-06	0.0	13	13	48	
0	2	3	1	2	2	0.0	0.0	0.0	0.0	-0.68E-06	0.0	14	19	25	EV
0	2	4	1	2	2	0.0	0.0	0.0	0.0	-0.68E-06	0.0	6	22	0	

(Columns a–ω+M are under the spanning header "Amplitude"; columns day, hr, min under "Period".)

Table 1 continued

s	n	p	q	h	k	a	e	i	Ω	ω	ω+M	day	hr	min	
(M	a/a*	M*	Ω-Ω*	ω	ω*)										
0	2	0	1	-2	0	0.0	-0.27E-06	0.0	0.0	-0.32E-05	0.0	24	1	3	
0	2	2	1	-2	2	0.0	0.10E-06	0.0	0.0	0.12E-05	0.0	8	13	3	
0	2	3	1	-2	2	0.0	0.0	0.0	0.0	0.18E-06	0.0	6	5	17	
0	2	-1	2	0	0	0.0	0.0	-0.28E-06	0.0	0.0	0.0	48	22	36	
0	2	0	2	0	0	0.0	0.0	0.27E-05	-0.33E-05	0.0	0.59E-05	42	22	49	
0	2	1	2	0	0	0.0	0.0	0.85E-07	0.0	0.0	0.0	14	22	13	
0	2	1	2	0	2	0.0	0.0	-0.29E-06	0.35E-06	0.0	-0.63E-06	18	4	12	
0	2	2	2	0	2	0.0	0.0	0.52E-05	-0.64E-05	0.0	0.12E-04	10	3	4	
0	2	2	2	0	2	0.0	0.0	-0.49E-07	0.0	0.0	0.0	6	0	10	VA
0	2	2	2	0	2	0.0	0.0	-0.26E-06	0.0	0.0	0.0	32	6	0	VA
0	2	3	2	0	2	0.0	0.0	0.77E-06	-0.94E-06	0.0	0.17E-05	7	0	28	
0	2	3	2	0	2	0.0	0.0	0.12E-06	-0.15E-06	0.0	0.27E-06	5	16	31	EV
0	2	3	2	0	2	0.0	0.0	0.20E-06	-0.24E-06	0.0	0.43E-06	9	3	57	EV
0	2	4	2	0	2	0.0	0.0	0.88E-07	-0.11E-06	0.0	0.19E-06	5	8	55	
0	2	4	2	0	2	0.0	0.0	0.29E-07	0.0	0.0	0.0	4	13	20	EV
0	2	4	2	0	2	0.0	0.0	0.42E-07	0.0	0.0	0.0	6	13	2	EV
0	2	-1	2	2	0	0.0	0.0	0.0	0.0	0.69E-06	0.0	19	20	22	
0	2	0	2	2	0	0.0	-0.48E-05	0.0	0.0	0.58E-04	0.0	149	23	56	
0	2	1	2	2	0	0.0	0.0	0.0	0.0	-0.94E-06	0.0	26	23	48	
0	2	1	2	2	2	0.0	-0.32E-06	0.0	0.0	0.38E-05	0.0	39	21	19	
0	2	2	2	2	2	0.0	0.39E-05	0.12E-06	0.19E-06	-0.46E-04	0.35E-06	14	12	54	
0	2	2	2	2	2	0.0	0.0	0.0	0.0	0.36E-06	0.0	7	7	48	VA
0	2	2	2	2	2	0.0	0.0	0.0	0.0	0.47E-04	0.0	945	2	47	VA
0	2	3	2	2	2	0.0	0.50E-06	0.0	0.0	-0.60E-05	0.0	8	21	20	
0	2	3	2	2	2	0.0	0.76E-07	0.0	0.0	-0.90E-06	0.0	6	20	34	EV
0	2	3	2	2	2	0.0	0.14E-06	0.0	0.0	-0.17E-05	0.0	12	15	9	EV
0	2	2	0	-2	2	0.0	-0.58E-07	0.0	0.0	-0.68E-06	0.0	9	11	42	
0	2	-2	1	0	-2	0.0	0.0	-0.29E-06	0.0	0.45E-06	-0.82E-06	15	16	6	

Table 1 continued

s	n	p	q	h	k	a	e	i	Ω	ω	ω+M	day	hr	min	
(M	a/a*	M*	Ω−Ω*	ω	ω*)										
0	2	-1	1	0	0	0.0	0.0	0.83E-06	0.33E-06	0.13E-06	-0.23E-05	31	4	16	
0	2	-1	1	0	0	0.0	0.0	0.41E-05	0.0	0.0	0.18E-04	793	17	8	EV
0	2	-1	1	0	0	0.0	0.0	-0.80E-07	0.0	0.0	-0.35E-06	15	6	56	EV
0	2	0	1	0	0	0.0	0.0	0.25E-04	-0.99E-05	-0.40E-04	0.71E-04	85	21	39	
0	2	0	1	0	0	0.0	0.0	0.0	0.0	0.0	-0.25E-06	12	14	23	VA
0	2	0	1	0	0	0.0	0.0	0.0	0.0	0.0	0.36E-06	17	19	55	VA
0	2	1	1	0	0	0.0	0.0	0.48E-06	-0.19E-06	-0.75E-06	0.14E-05	18	1	33	
0	2	1	1	0	0	0.0	0.0	0.59E-07	0.0	0.0	0.26E-06	11	6	36	EV
0	2	1	1	0	0	0.0	0.0	0.24E-06	0.0	0.0	0.10E-05	45	9	47	EV
0	2	1	1	0	2	0.0	0.0	0.22E-06	0.0	0.0	0.96E-06	23	1	16	
0	2	2	1	0	2	0.0	0.0	-0.35E-05	0.14E-05	0.55E-05	-0.10E-04	11	11	34	
0	2	2	1	0	2	0.0	0.0	0.31E-07	0.0	0.0	0.14E-06	6	11	1	VA
0	2	2	1	0	2	0.0	0.0	0.0	0.0	0.0	0.11E-05	51	15	17	VA
0	2	3	1	0	2	0.0	0.0	-0.50E-06	0.20E-06	0.79E-06	0.14E-05	7	15	28	
0	2	3	1	0	2	0.0	0.0	-0.78E-07	0.0	0.12E-06	-0.22E-06	6	2	12	EV
0	2	3	1	0	2	0.0	0.0	-0.13E-06	0.0	0.21E-06	0.37E-06	10	6	13	EV
0	2	4	1	0	2	0.0	0.0	-0.56E-07	0.0	0.0	-0.24E-06	5	17	31	
0	2	4	1	0	2	0.0	0.0	0.0	0.0	0.0	-0.81E-07	4	19	27	EV
0	2	-2	1	2	-2	0.0	0.60E-07	0.0	0.0	-0.70E-06	0.0	10	15	58	
0	2	-1	1	2	0	0.0	0.13E-06	0.0	0.0	-0.15E-05	0.0	16	2	57	
0	2	0	1	2	0	0.0	0.49E-05	0.0	0.0	-0.57E-04	0.0	54	14	54	
0	2	1	1	2	0	0.0	-0.32E-06	0.0	0.0	0.37E-05	0.0	39	8	38	
0	2	1	1	2	2	0.0	0.0	0.0	0.0	0.25E-05	0.0	74	11	12	
0	2	2	1	2	2	0.0	0.16E-05	0.0	0.0	-0.19E-04	0.0	17	11	59	
0	2	3	1	2	2	0.0	0.20E-06	0.0	0.0	-0.23E-05	0.0	9	21	57	
0	2	3	1	2	2	0.0	0.0	0.0	0.0	-0.34E-06	0.0	7	10	50	EV
0	2	4	2	2	2	0.0	0.54E-07	0.0	0.0	-0.64E-06	0.0	6	9	38	

Table 1 continued

s (M	n a/a*	p M*	q Ω−Ω*	h ω	k ω*)	a	e	i	Ω	ω	ω+M	day	hr	min	
0	2	4	2	2	2	0.0	0.0	0.0	0.0	-0.21E-06	0.0	5	6	37	EV
0	2	4	2	2	2	0.0	0.0	0.0	0.0	-0.32E-06	0.0	8	3	18	EV
0	2	2	2	-2	2	0.0	0.0	0.0	0.0	-0.39E-06	0.0	7	18	30	
1	2	0	0	2	0	-0.18E-04	0.14E-07	-0.30E-08	-0.46E-08	-0.16E-06	0.0	0	1	49	
1	2	0	1	0	0	-0.18E-04	-0.14E-07	0.19E-08	-0.74E-09	-0.17E-06	0.0	0	1	50	
1	2	2	1	0	2	0.19E-04	0.15E-07	-0.20E-08	0.78E-09	0.17E-06	0.0	0	1	50	
1	2	0	1	2	0	0.62E-04	-0.48E-07	0.35E-08	0.50E-08	0.56E-06	0.0	0	1	50	
1	2	2	1	2	2	-0.65E-04	0.51E-07	-0.37E-08	-0.53E-08	-0.59E-06	0.0	0	1	50	
1	2	3	1	2	2	-0.14E-04	0.11E-07	-0.79E-09	-0.11E-08	-0.13E-06	0.0	0	1	51	
1	2	2	2	0	2	-0.16E-04	-0.12E-07	0.33E-08	0.40E-08	-0.14E-06	0.0	0	1	51	
1	2	0	2	2	0	-0.23E-04	0.18E-07	0.11E-08	0.17E-08	-0.21E-06	0.0	0	1	50	
1	2	2	2	2	2	-0.19E-03	0.15E-06	0.88E-08	0.14E-07	-0.17E-05	0.0	0	1	50	
1	2	3	2	2	2	-0.40E-04	0.31E-07	0.19E-08	0.30E-08	-0.36E-06	0.0	0	1	51	
2	2	2	2	2	2	0.72E-03	-0.20E-08	-0.17E-07	-0.27E-07	-0.10E-06	0.0	0	0	55	
0	3	-1	0	1	-1	0.0	0.0	0.0	0.0	0.83E-06	0.0	18	23	24	
0	3	1	0	1	1	0.0	0.0	0.0	0.0	0.19E-05	0.0	43	22	53	
0	3	-1	0	-1	-1	0.0	0.0	0.0	0.0	-0.19E-05	0.0	43	22	53	
0	3	1	0	-1	1	0.0	0.0	0.0	0.0	-0.83E-06	0.0	18	23	24	
0	3	-1	1	-1	-1	0.0	0.0	0.0	0.0	-0.30E-05	0.0	90	0	13	
0	3	1	1	-1	1	0.0	0.0	0.0	0.0	0.33E-06	0.0	15	13	0	
0	3	3	1	-1	3	0.0	0.0	0.0	0.0	0.21E-06	0.0	7	3	40	
0	3	1	2	1	1	0.0	-0.16E-06	0.0	0.0	0.20E-05	0.0	21	17	21	
0	3	3	2	1	3	0.0	0.69E-07	0.0	0.0	-0.83E-06	0.0	8	5	33	
0	3	4	2	1	3	0.0	0.0	0.0	0.0	-0.19E-06	0.0	6	1	16	
0	3	1	2	-1	1	0.0	0.0	0.0	0.0	0.37E-06	0.0	13	3	51	
0	3	3	2	-1	3	0.0	0.0	0.0	0.0	-0.21E-06	0.0	6	14	29	
0	3	3	3	1	3	0.0	-0.62E-07	0.0	0.0	0.73E-06	0.0	7	12	16	
0	3	4	3	1	3	0.0	0.0	0.0	0.0	0.17E-06	0.0	5	15	42	

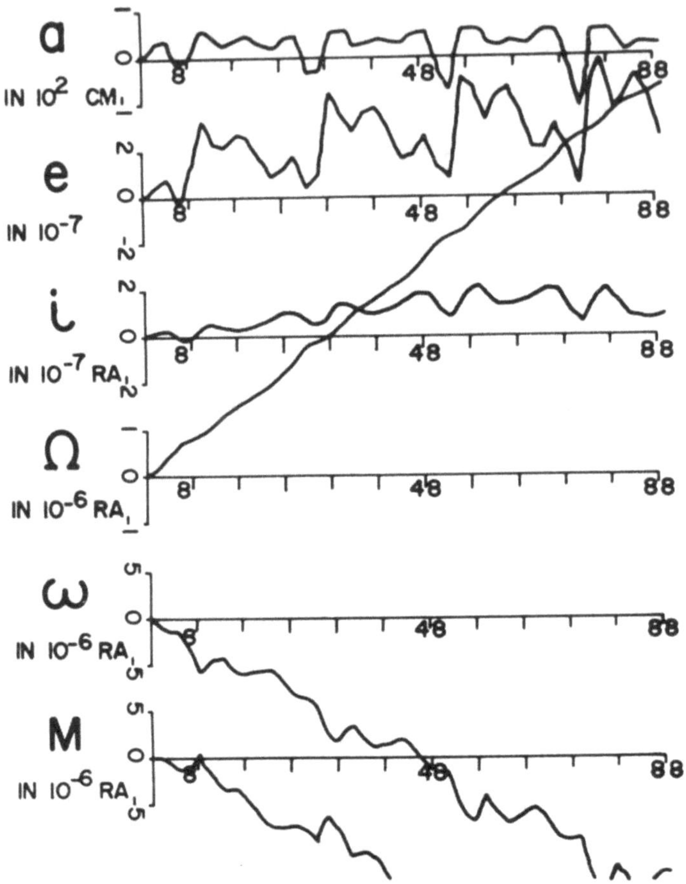

Figure 1. Difference between analytical theory and numerical integration for simultaneous terrestrial and lunar influences. Coupling is accomplished by secular terms. Abscissa: time in hours.

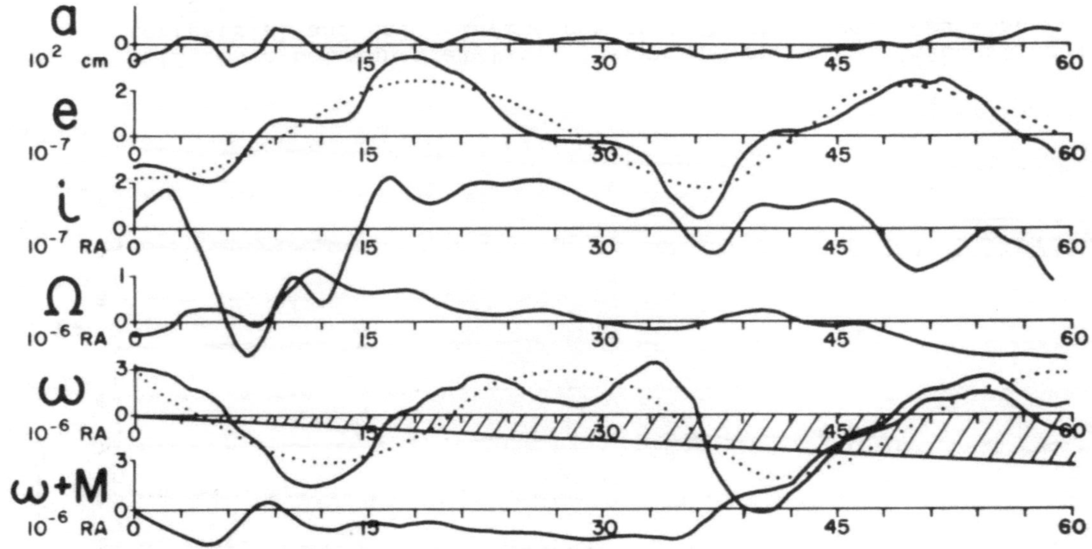

Figure 2. Difference between analytical theory and numerical integration for terrestrial and lunar influences on D1C satellite. Abscissa: time in days.

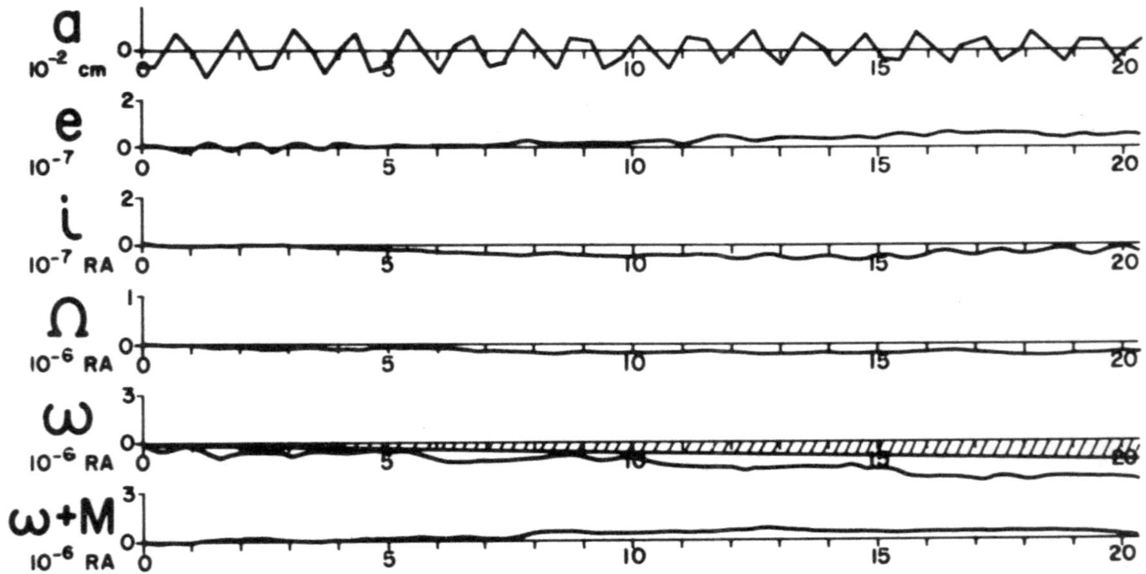

Figure 3. Difference between analytical theory and numerical integration for terrestrial and solar influences on D1C satellite. Abscissa: time in days.

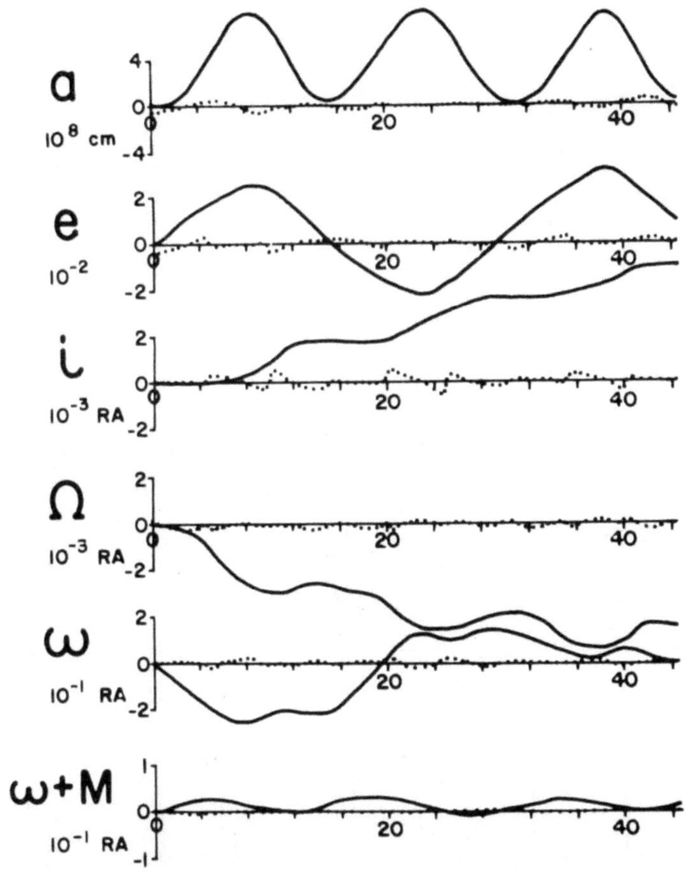

Figure 4. The elements are expressed in the equatorial 1950.0 system. Abscissa: time in days. ___ variation of the lunar elements with time (for ω+M the secular effect has been removed). ... difference between the real lunar motion and the lunar motion considered.

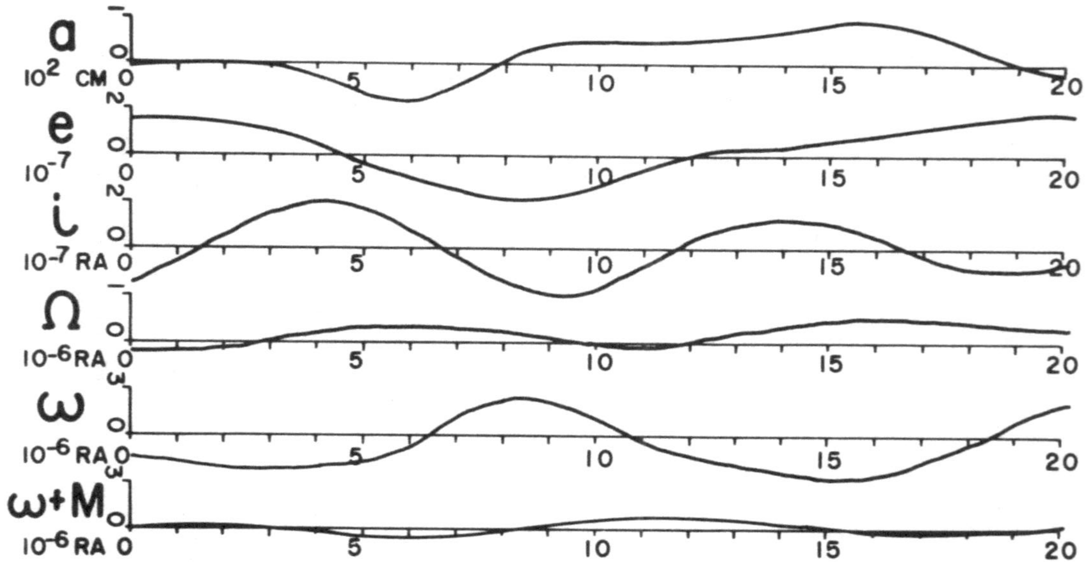

Figure 5. Difference between Remora method and numerical integration for simultaneous terrestrial and lunar influences. Abscissa: time in days.

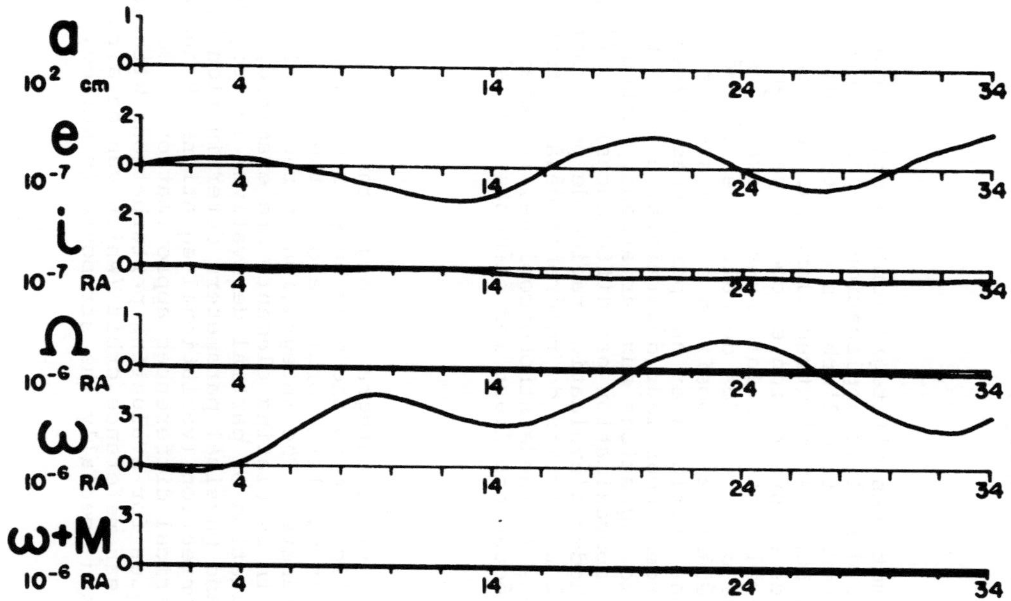

Figure 6. Coupling effect by short period terms between Earth and Moon: Difference between Remora theory with and without short period terms. Abscissa: time in days.

Numerical Methods of Orbital Dynamics

Georges BALMINO
Groupe de Recherches de Géodésie Spatiale
C.N.E.S., Bretigny S/Orge, France

Résumé

Différentes méthodes d'integration numérique sont passées en revue dans leur application au problème de n-corps, la détermination d'orbites de satellites de la Terre, de la Lune, et d'autres planètes. De même que les techniques d'intégration en coordonnées rectangulaires et en éléments osculateurs sont examinées, nous décrivons quelques méthodes de calcul des dérivées partielles des éléments de position et vitesse par rapport aux conditions initiales et aux paramètres physiques du modèle de forces, pour leur détermination par correction différentielle (équations aux variations, méthode de Peano Baker, approximation par différences). Enfin, quelques méthodes d'évaluation rapide de perturbations par rapport à une orbite de référence déjà perturbée, sont comparées (méthode stroboscopique, intégration numérique des équations liant les éléments moyens).

Abstract

Different numerical integration algorithms are reviewed in their application to the many body problem, the determination of satellite orbits around the Earth, Moon and other planets. The integration in rectangular coordinates and in osculating elements are examined, as well as the computation of partial derivatives in trajectory estimation and physical parameters determination by differential correction (variational equations, Peano Baker method, numerical differences approximation). Besides, fast techniques for evaluating perturbations with respect to a mean reference orbit (which can be already disturbed) and specially the stroboscopic method, are described.

Introduction

Bien des progrès ont été accomplis dans le développement des méthods numériques de résolution des systèmes d'-equations différentielles. Au contraire ce n'est que récemment que des disciplines comme la mécanique céleste et la géodésie spatiale ont bénéficié de l'existence de ces nouvelles techniques.

Nous ne considérons ici que les problèmes à conditions initiales traduisant la loi fondamentale de la mécanique dans un certain système de référence; c'est-à-dire que nous nous intéressons, d'une part, au calcul des trajectoires d'un corps lorsque le champ de force, et la position et la vitesse de ce corps sont connues à un instant origine, d'autre, à la détermination de cés mêmes conditions initiales ainsi que des paramètres physiques décrivant le champ de forces à l'aide d'observations de quantité qui sont fonctions de ces paramètres. Ces différentes études peuvent être effectuées à l'aide d'une théorie analytique ou d'une intégration numérique du mouvement. Au contraire des méthodes analytiques dont le développement, de plus en plus précis, est maintenant effectué à l'aide de languages algebriques sur ordinateurs (voir, par exemple: Kovalevsky, (1)), les méthodes numériques reviennent toutes à écrire les équations différentielles du mouvement en coordonnées rectangulaires ou en fonction des éléments osculateurs (équations de Lagrange ou de Gauss) et à les intégrer par une méthode de calcul pas à pas. Les techniques sont nombreuses et variées et nous distinguons les méthodes à pas séparés, à pas liés, à pas fixe ou variable, les méthodes de prédiction-correction, les récentes méthodes d'extrapolation par itération, etc... Par ailleurs, remarquant que ces méthodes sont basées sur l'approximation polynomiale de la trajectoire et donc n'intègrent exactement que des polynômes d'un certain degré, divers auteurs ont tenté avec succès des méthodes dérivées qui intègrent exactement le mouvement Képlérien (par exemple).

La détermination des conditions initiales et des paramètres décrivant le champ des forces responsables de la trajectoire, ceci à partir d'observations, requiert le calcul des dérivées partielles de quantités mesurées, et finalement des coordonnées et vitesse du corps, par rapport aux inconnues sus-nommées. A côte des méthodes expéditives telles que: l'approximation des dérivées partielles par différences finies (souvent coûteuse et relativement lourde), leur calcul par intégration numérique d'équations aux variations (technique qui reste encore la plus précise), d'autres ont vu le jour. Par exemple, l'intégration des systèmes variationnels par la méthode de Peano-Baker, l'utilisation d'intégrales premières pour les équations linéaires, enfin l'utilisation particulière du noyau résolvant du système homogène déterminant l'orbite de référence.

Enfin, un dernier aspect de la détermination des orbites est celui des techniques qui moyennent les perturbations pendant un certain intervalle de temps (en général une révolution du mouvement osculateur où d'un mouvement perturbé de référence). Les bases théoriques de ces méthodes sont élémentaires, ce qui est plus difficile est leur mise en pratique et leur utilisation dans divers problèmes de mécanique orbitale.

Calcul des Trajectoires. Methodes D'Integration Numerique Les Plus Courantes

Dans le calcul d'une trajectoire pas à pas, il faut établir:

1. Un formulaire permettant de calculer l'approximation suivante;

2. Le formalisme d'estimation des erreurs de troncature par pas pour cette approximation;
3. Des critères permettant de décider à quel moment il convient d'accepter le résultat, eu égard à la précision cherchée.
4. Une stratégie pour, si nécessaire, déterminer à quel instant il faut changer le pas et calculer le meilleur pas possible, en fonction des seconds membres du système différentiel étudié et des résultats et pas précédents.

Une méthode d'intégration numérique comporte toujours ces quatre points, bien que les différentes variantes utilisées et les mises en oeuvre pratiques se passent souvent du dernier (d'ailleurs le plus compliqué). Pour simplifier, divers auteurs, par exemple Danby (2), utilisent des méthodes d'intégration autour d'une orbite déjà perturbée ou de l'orbite Képlérienne régularisée et bénéficient de formulaires que sont rigoureux pour le mouvement de référence. Suivant la même idée il est possible d'intégrer le mouvement perturbé autour d'une trajectoire képlérienne par la méthode de Picard d'approximations successives; en particulier, Broucke (3) a mis au point une variante très rapide par séries de Chebychev.

En ce qui concerne le formalisme de calcul d'une approximation donnée, on distingue généralement les méthodes à pas séparés comme celles de Runge-Kutta, qui donnent de bons résultats, mais n'utilisent pas l'information fournie par les points déjà calculés, autre que le dernier, et les méthodes à pas liés: en effet, en raison de la régularité habituelle des courbes, on peut obtenir à peu de frais des renseignements que les méthodes à pas séparés ne fournissent qu'avec un volume de calcul considérable et au prix d'une mise en oeuvre laborieuse; les méthodes à pas liés, pour lesquelles donc les valeurs trouvées à $t > t_i$ ne dépendent non seulement des valeurs à $t = t_i$, mais aussi à $t = t_{i-1}$, t_{i-2}... ont l'inconvénient de ne pas respecter la propriété fondamentale de la solution d'un problème de conditions initiales et cette incorrection se paiera par des phénéomènes parasites gênants tels que la difficulté de faire varier la longueur du pas au cours de l'intégration, la nécessité de formules de départ (démarrage), et parfois l'instabilité (4).

L'estimation des erreurs de troncature peut se faire de la manière suivante, ainsi décrite par Gorn et Moore (5) et utilisée par exemple par Roy et al. (6), et Balmino (7): soit u(t) la solution du système différentiel sous forme canonique

$$\frac{dy}{dt} = f(t,y) \quad , \quad y(t_0) = y_0 \quad , \tag{1}$$

obtenue par une quelconque méthode numérique. En négligeant les erreurs d'arrondis, l'accumulation des erreurs de troncature e (t) est:

$$e(t) = u(t) - y(t) \quad . \tag{2}$$

On définit alors l'erreur locale de troncature comme étant l'erreur après un pas de longueur h, i.e.

$$e_\ell = e(t+h) - e(t) \quad . \tag{3}$$

Partant donc du point (t,u) et obtenant, à l'aide de l'algorithme utilisé, une approximation de y(t+h) d'abord avec le pas h, puis en deux étapes avec le pas h/2, δ étant la différence entre ces deux approximations et p l'ordre de la méthode (le rang maximum des différences successives de f considérées dans l'algorithme), nous avons:

$$e_\ell = \frac{\delta}{1 - 1/2\ p} + 0\left(h^{p+2}\right) \quad . \tag{4}$$

Pour la plupart des méthodes utilisées, $p \geq 5$ et e_ℓ est proche de $\delta + O(h^{p+2})$. Cette méthode est évidemment précise mais lourde et onéreuse. Les méthodes telles que celles utilisant des séries de Taylor donnent par contre immédiatement cette erreur de troncature par pas, en calculant le terme suivant (et négligé) de chaque développement en série.

Le choix du pas h est assez délicat. D'une part, il ne doit pas être trop petit pour éviter d'allonger démesurément les calculs et de trop accumuler les erreurs d'arrondi; on démontre, en effet, que si l'on multiplie le nombre de pas par un facteur α, les erreurs d'arrondi sont multipliées par $\alpha^{1/2}$ pour les méthodes par séries de Taylor (d'après Henrici (8), Black (9)), et par $\alpha^{3/2}$ pour les méthodes du type Cowell (10,11).

D'autre part, un pas trop grand donnerait des différences d'ordre élévé, incompatibles avec l'hypothèse que l'on a faite en négligeant les différences d'ordre supérieur à un ordre donné. Par exemple, dans le problème de deux corps perturbé et pour des orbites très excentriques, la force varie beaucoup avec le temps et un pas qui conviendrait à l'apogée risque d'être trop long au voisinage du périgée. Il est alors nécessaire de prévoir une procédure de changement de pas. Pour les méthodes de type Runge-Kutta, variantes de Nyström (12, 13), et à pas séparé, il est aisé de modifier le pas suivant h^* lorsque celui-ci a été calculé à partir d'informations recueillies à l'étape précédente de pas h. On démontre (14) que:

$$h^* \simeq 0.9h\left(\frac{\varepsilon}{\tau}\right)^{1/p} \tag{5}$$

où ε est la précision requise, et τ mesure le maximum de l'erreur de troncature (peut être estimé par e_ℓ). Cette expression de h^* suppose que les dérivées d'ordres les plus élevés, considérées, ne varient que très peu entre $t + h$ et $t+ h^*$--ce qui est rarement le cas dans les problèmes à n corps et en géodésie spatiale où la fonction f est très compliquès et prèsente d'importantes variations à très courte période. C'est afin de compenser cette hypothèse faible que les auteurs ont introduit le facteur 0.9, déterminé d'ailleurs expérimentalement.

Pour les méthodes à pas liés, de type Cowell, la modification quelconque du pas est difficile à mettre en pratique et l'on se contente de multiplier ou de diviser le pas par deux, opérations pour lequelles les formules de reconstitution des tableaux de progression sont plus simples. La décision de changer le pas se fait soit en calculant les différences d'ordre p+1 et en examinant les variations de celles d'ordre p-1 (pour une méthode d'ordre p), mais cela conduit souvent à des modifications intempestives du pas, et finalement à l'instabilité, soit à l'aide de critères plus empiriques mais très efficaces: par exemple pour un satellite de la Terre, l'intégration numérique à l'aide de deux pas h et h/2 pendant une révolution et la reprise avec les changements de pas ad'hoc suivant les erreurs locales de troncature e_ℓ, les changements ultérieurs (pour les révolutions suivantes) ayant lieu pour des positions semblables du corps sur son orbite (testées suivant la valeur de l'anomalie vraie ou la distance géocentrique).

Nous allons maintenant passer en revue quelques méthodes d'intégration numérique parmis les meilleures et les plus utilisées.

Méthodes de Runge-Kutta

Ce sont des méthodes à pas séparés. Appelant y_n la valeur de la solution de l'équation (1) à l'instant $t_n = t_0 + nh$ et y_{n+1} la valeur à estimer en t_{n+1}, toutes les méthodes explicites du type Runge-Kutta équivalent à écrire:

$$y_{n+1} = y_n + \sum_{i=1}^{\nu} w_i k_i \qquad (6)$$

où ν est le nombre d'évaluations des dérivées, w_i des poids, et k_i des fonctions vérifiant les formules de récurrence ci-après (15,16):

$$k_i = hf(t_n + c_i h,\ y_n + \sum_{j=1}^{\nu} a_{ij} k_j)$$

$$c_1 = 0 \quad ; \quad i = 1,2,\ldots\nu \quad . \qquad (7)$$

Les différences variantes de Runge-Kutta sont définies par différentes choix des paramètres c_i, a_{ij}, w_i et divers critères de convergence et stabilité. Ces formules peuvent se représenter symboliquement par le tableau:

$$\begin{array}{c|cccc} c_1 & a_{11} & a_{12} & \cdots & a_{1\nu} \\ c_2 & a_{21} & a_{22} & \cdots & a_{2\nu} \\ \vdots & \vdots & \vdots & & \vdots \\ c_\nu & a_{\nu 1} & a_{\nu 2} & \cdots & a_{\nu\nu} \\ \hline & w_1 & w_2 & \cdots & w_\nu \end{array} \qquad (8)$$

Les tables 1 à 3 donnent les tableaux correspondants respectivement aux variantes de Kutta (4è ordre), Butcher (6è ordre) (17), et Shanks (8è ordre) (18).

Méthodes de récurrence (Séries de Taylor)

De très nombreuses variantes de ces méthodes existent.

Retenons particulièrement l'une d'entre elles, mise en oeuvre par Roy, Moran et Black (6,9,15,19). Pour le problème des deux corps, formulé par:

$$\ddot{\bar{r}} = -\mu\, \bar{r}/r^3 \tag{9}$$

où r est la distance des deux corps, $\bar{r}$ le rayon vecteur correspondant, μ une constante positive, les auteurs posent:

$$u = 1/r^3 \qquad s = \bar{r}\cdot\dot{\bar{r}}$$

$$w = 1/r^2 \qquad \sigma = ws \tag{10}$$

Développant ensuite chaque paramètre Q en série de Taylor

$$Q = \sum_{i=0}^{\infty} Q_i t_i \tag{11}$$

on arrive aux formules de récurrence ci-après:

$$u_j = (3/j) \sum_{i=0}^{j-1} \sigma_{j-i-1} u_i$$

$$w_j = (2/j) \sum_{i=0}^{j-1} \sigma_{j-i-1} w_i$$

$$s_j = \sum_{i=0}^{j} (i+1)\, \bar{r}_{i+1} \bar{r}_{j-i}$$

$$\sigma_j = \sum_{i=0}^{j} w_i s_{j-i}$$

$$\bar{r}_{j+2} = -\frac{\mu}{(j+1)(j+2)} \sum_{i=0}^{j} \bar{r}_i\, u_{j-i} \quad . \tag{12}$$

Moran (19) a appliqué cette technique au problème restreint des trois corps, P étant de masse négligeable en orbite autour du corps J de masse m, de mouvement perturbé par un troisième corps S de masse M. Moran applique alors la méthode des séries de Taylor à l'equation de Encke obtenue en considérant les perturbations par rapport à une orbite képlérienne (en l'absence du corps perturbateur S).

Méthodes de Cowell

Les algorithmes, à pas liés, de résolution d'un système d'équations

$$\ddot{x} = f(x,\dot{x},t)$$

$$\begin{cases} x(t_0) = x_0 \\ \dot{x}(t_0) = \dot{x}_0 \end{cases} \qquad (13)$$

sont tous de la forme

$$x_{n+1}^{(1)} + \sum_{p=1}^{P} A_p\, x_{n+1-p} = h^2 \sum_{p=1}^{N} B_p f_{n+1-p}$$

$$\dot{x}_{n+1}^{(1)} + \sum_{p=1}^{P-1} A'_p\, \dot{x}_{n+1-p} = h \sum_{p=1}^{N-1} B'_p\, f_{n+1-p} \qquad (14)$$

pour la partie prédiction, et

$$x_{n+1} + \sum_{p=1}^{M} A_p^*\, x_{n+1-p}$$

$$= h^2\left\{\sum_{p=1}^{R} B_p^*\, f_{n+1-p} + B_0^*\, f\left[x_{n+1}^{(1)}, \dot{x}_{n+1}^{(1)}, t_{n+1}\right]\right\}$$

$$\dot{x}_{n+1} + \sum_{p=1}^{M-1} A_p^{*\prime}\, \dot{x}_{n+1-p}$$

$$= h\left\{\sum_{p=1}^{R-1} B_p^{*\prime}\, f_{n+1-p} + B_0^{*\prime}\, f\left[x_{n+1}^{(1)}, \dot{x}_{n+1}^{(1)}, t_{n+1}\right]\right\}$$

$$(15)$$

pour la partie correction. Précisément, les formules (14) prédisent une valeur $\left\{x_{n+1}^{(1)}, \dot{x}_{n+1}^{(1)}\right\}$, qui est corrigée par formule (15); cette étape de correction est en général itérée en y substituant la nouvelle valeur $(x_{n+1}, \dot{x}_{n+1})$ à la place de $\left\{x_{n+1}^{(1)}, \dot{x}_{n+1}^{(1)}\right\}$. Dans ces expressions nous avons posé naturellement

$$t_n = t_0+nh \quad , \quad x_n = x(t_n) \quad ,$$

$$\dot{x}_n = \dot{x}(t_n) \quad , \quad f_n = f(x_n,\dot{x}_n,t_n) \quad ,$$

h est le pas d'intégration et A_p, B_p, A_p', B_p', A_p^*, B_p^*, $A_p^{*\prime}$, $B_p^{*\prime}$, des constantes dépendent de l'algorithme utilisé.

Les méthodes les plus classiques se limitent à $P = M = 2$, $A_2 = A_1' = A_2^* = A_1^{*\prime} = 1$, et $A_1 = A_1^* = -2$. Par exemple les formules de prédiction de Störmer sont définies pour $R = \nu$, $N = \nu+\Lambda$ et des coefficients B_p choisis de façon à intégrer rigoureusement l'equation (13) lorsque f est un polynome en t de degré ν (20). De même les formules de correction (pour $P = 2$, $A_2 = A_1' =$ etc...) où l'on pose $x_{n+1}^{(1)} \equiv x_{n+1}$ et $\dot{x}_{n+1}^{(1)} \equiv \dot{x}_{n+1}$ de manière à définir implicitement $(x_{n+1}, \dot{x}_{n+1})$, se réduisent aux classiques formules de Cowell (voir par exemple Henrici (8), Ceschino and Kuntzmann(4)), pour lesquelles de très nombreuses variantes existent, comme celle de Kulikov (21),décrite par Kovalevsky (11), et utilisée par Balmino (22), Chassaing and Roumiguier (23), Moynot (24).

Nous nous limitons, ici, à la description d'un algorithme mixte (Adams et Cowell), mis au point par Hubbard et Broadwater et utilisé en particulier par Cohen et Oesterwinter (25).

Les auteurs calculent tout d'abord les coefficients a_j et c_j donnés par la Table 4.

Ces coefficients permettent de mettre en oeuvre une méthode d'ordre variable, compris entre 3 et 16, tant du point de vue du démarrage que de la progression. On définit tout d'abord une série de coefficients auxiliaires par les formules (16):

$$\gamma'_j = \sum_{i=0}^{\inf(j+1,b)} (-1)^i \binom{b}{i} (a_{j+1-i} - c_{j+1-i}) \qquad j = 0,1,2,\ldots a+b$$

$$\gamma_i = (-1)^i \sum_{j=i}^{a+b} \binom{j}{i} \gamma'_j \qquad i = 0,1,2,\ldots a+b$$

$$b_j = (-1)^j \binom{a+b+1}{j+1} \qquad j = 0,1,2,\ldots a+b$$

$$\alpha^*_j = (-1)^j \sum_{i=j}^{c} \binom{i}{j} c_i \qquad j = 0,1,2,\ldots c$$

$$\beta^*_j = (-1)^j \sum_{i=j}^{c} \binom{i}{j} a_i \qquad j = 0,1,2,\ldots c$$

$$c'_j = \sum_{j=0}^{i} c_j \qquad i = 0,1,2,\ldots m$$

$$\alpha'_j = (-1)^j \sum_{i=j}^{m} \binom{i}{j} c'_i \qquad j = 0,1,2,\ldots m$$

$$a'_i = \sum_{j=0}^{i} a_j \qquad i = 0,1,2,\ldots m$$

$$\begin{Bmatrix} \beta'_j \\ \alpha_j \\ \beta_j \end{Bmatrix} = (-1)^j \sum_{i=j}^{m} \binom{i}{j} \begin{Bmatrix} a'_i \\ c_i \\ a_i \end{Bmatrix} \quad ; \; j = 0,1,2,\ldots m \tag{16}$$

Dans ces formules, $\binom{\mu}{\nu}$ désigne un coefficient du binôme C_μ^ν, c et m sont les ordres des algorithmes de démarrage et de démarrage et de progression, respectivement; le rôle de a et b sera expliqué ci-dessous. Nous intégrons le système (formule 13), où $x, \dot{x}, f$ sont des vecteurs.

Phase de démarrage

De x_0 et $\dot{x}_0$, on déduit $\ddot{x}_0 = f(x_0, \dot{x}_0, t_0)$. Le tableau de démarrage est initialisé à: $\ddot{x}_n = \ddot{x}_0$, pour $-a \leq n \leq b$. Les schéma ci-après, résume le processus, avec $a \leq c$ et $b \leq c$:

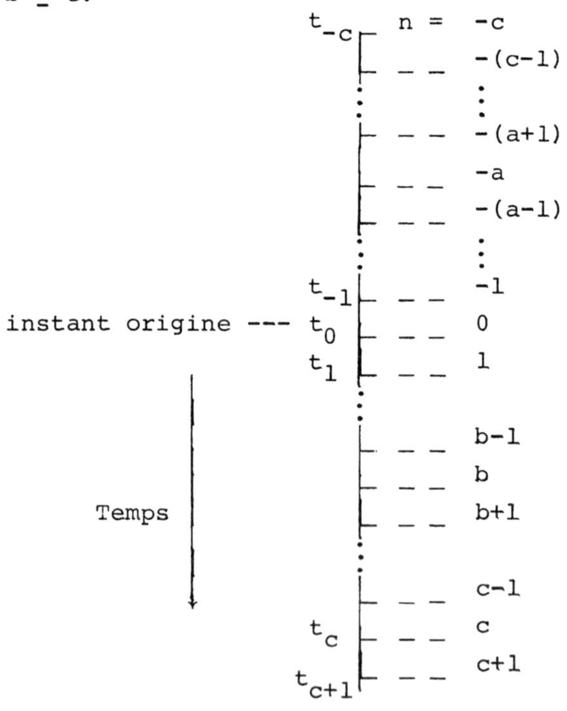

h étant le pas d'intégration, nous avons ensuite

$$x_{-1} = x_0 - h\dot{x}_0 + h^2 \sum_{i=0}^{a+b} \gamma_i \ddot{x}_{b-i} \tag{17}$$

Puis on extrapole les accélérations $\ddot{x}$ au delà de -a et b, sur c intervalles, par

$$\ddot{x}_{-n} = \sum_{j=0}^{a+b} b_j \ddot{x}_{-n+1+j} \quad ; \quad n = a+1, a+2, \ldots c$$

$$\ddot{x}_{+n} = \sum_{j=0}^{a+b} b_j \ddot{x}_{n-1-j} \quad ; \quad n = b+1, b+2, \ldots c. \tag{18}$$

La position et la vitesse sont ensuite déterminées, entre n = -a et n = b, par

$$x_n = 2x_{n-1} - x_{n-2} + h^2 \sum_{j=0}^{c} \alpha_j^* \ddot{x}_{n-j} \qquad (n = 1, 2, \ldots b)$$

$$x_{-n} = 2x_{-n+1} - x_{-n+2} + h^2 \sum_{j=0}^{c} \alpha_j^* \ddot{x}_{-n+j} \qquad (n = 2, 3, \ldots a)$$

$$\dot{x}_n = \dot{x}_{n-1} + h \sum_{j=0}^{c} \beta_j^* \ddot{x}_{n-j} \qquad (n = 1, 2, \ldots b)$$

$$\dot{x}_{-n} = \dot{x}_{-n+1} - h \sum_{j=0}^{c} \beta_j^* \ddot{x}_{-n+j} \qquad (n = 1, 2, \ldots a) . \tag{19}$$

On calcule enfin: $\ddot{x}_n = f(x_n, \dot{x}_n, t_n)$ pour $-a \leq n \leq b$.

Phase de progression

Celle-ci progresse pas à pas. L'approximation au point suivant est effectuée par

$$x_n = 2x_{n-1} - x_{n-2} + h^2 \sum_{j=0}^{m} \alpha'_j \ddot{x}_{n-1-j}$$

$$\dot{x}_n = \dot{x}_{n-1} + h \sum_{j=0}^{m} \beta'_j \ddot{x}_{n-1-j}$$

$$\ddot{x}_n = f(x_n, \dot{x}_n, t_n) \qquad (20)$$

où m est l'ordre des formules de prédiction (20), en général différent de c, ($m \leq c$). Les composantes x et $\dot{x}$ sont enfin corrigées à l'aide des formules de correction (formule 21)

$$x_n = 2x_{n-1} - x_{n-2} + h^2 \sum_{j=0}^{m} \alpha_j \ddot{x}_{n-j}$$

$$\dot{x}_n = \dot{x}_{n-1} + h \sum_{j=0}^{m} \beta_j \ddot{x}_{n-j} \qquad (21)$$

et les accélérations $\ddot{x}_n$ sont à nouveau corrigées en conséquence. Le processus prédiction-correction peut alors être itéré L fois. Par exemple, pour l'intégration numérique du mouvement des 9 grosses planètes du système solaire, Oesterwinter et Cohen ont utilisé: $h = 0.4$ jour, $c = 14$, $a = b = 7$, $m = 12$, $K = 4$, $L = 0$ ou 1.

Pour un bon usage de ces méthodes, nous recommandons d'une part, d'utiliser des formules de démarrage d'ordre c au moins égal à $m + 1$, et d'autre part, de travailler en double précision dès que $m \geq 10$ afin de minimiser les erreurs d'arrondi.

Méthodes d'extrapolation

La théorie de l'extrapolation fut exposée par Bulirsch et Stoer en 1965 (26). Elle est à la base d'une des méthodes en précision numérique les plus puissantes qui soient en précision et en stabilité, mise en oeuvre par Bulirsch et Stoer (27), et utilisée en routine pour de nombreux travaux de prédiction d'orbite à long terme de géodésie ou en planétologie (7).

En effet, soit l'equation dans R^n: $dy/dx = f(x,y)$ avec x_0, $y_0 = y(x_0)$, où l'on suppose f de classe C^{k+1} dans le domaine étudié. On définira un algorithme de correction-prediction (sans phase de démarrage) de la manière suivante: la prédiction sera à pas multiple, i.e. fournira plusieurs valeurs possibles $P(h_s,x)$ de $y(x_0+H)$ suivant le pas h_s utilisé; la correction donnera la meilleure estimation de y en (x_0+H) d'après les différentes valeurs $P(h_s,x)$ en extrapolant ces valeurs à l'aide d'une fraction rationnelle $C_m = \left(C_m^{(i)}\right)_{i=1,2,\ldots n}$ - vecteur R^n, pour $h = 0$.

Algorithme de prédiction

Posons $x = x_0 + H = x_0 + \ell_s h_s$, $(\ell_s \varepsilon N)$. On définit alors $P(h_s,x) = P(h_s)$ comme suit,

$$x_{i+1} = x_i + h_s/2 \qquad i = 0,1,\ldots,2\ell_{s-1}$$

$$(\text{i.e. } x_{2\ell s} = x)$$

$$\eta_0 = y_0$$

$$\eta_1 = (x_1,h_s/2) = y_0 + f(x_0,y_0)\cdot\frac{h_s}{2}$$

$$\eta_2 = \eta(x_2,h_s/2) = \eta_0 + h_s f(x_1,\eta_1)$$

$$\eta_{i+1} = \eta\left(x_{i+1},\frac{h_s}{2}\right) = \eta_{i-1} + h_s f(x_i,\eta_i) \qquad (22)$$

(méthode dite "du point-milieu"),...et enfin, d'après les hypothèse sur f

$$P(h_s) = \frac{1}{2}\left[\eta_{2\ell s} + \frac{h_s}{2} f(x_{2\ell s},\eta_{2\ell s}) + \eta_{2\ell s-1}\right] . \tag{23}$$

On effectue cette opération m + 1 fois à l'aide de pas $h_j, h_{j+1},\ldots h_{j+m}$ extraits de la suite $\{h_\nu\}_\nu$ décroissante tendant vers 0

$$h_0 = H \quad , \quad h_1 = H/2 \quad , \quad h_2 = H/3 \quad ,$$

$$h_{\nu+1} = \frac{1}{2} h_{\nu-1} (\nu>2) \quad , \tag{24}$$

pour laquelle la sensibilité de la méthode aux erreurs d'arrondis est minimum (28).

Algorithme de correction

Soit donc $h_i, h_{i+1},\ldots h_{i+m}$, m+1 valeurs extraites de la suite (formule 24).

Considérons la fraction rationnelle $\mathcal{C}_m^{(i)}(h)$, définie par

$$\mathcal{C}_m^{(i)}(h) = \frac{\sum_{r=0}^{\mu} p_r^{(i)} h^{2r}}{\sum_{t=0}^{\lambda} q_t^{(i)} h^{2t}} \tag{25}$$

avec $\mu = (m/2)$, $\lambda = m-\mu$, l'un des coefficients $p_{r_0}^{(i)}$ n'étant pas nul. Les (m+1) coefficients

$p_r^{(i)}/p_{r_0}^{(i)}$, $q_t^{(i)}/p_{r_0}^{(i)}$, $(r = 0,\ldots\mu,\ r \neq r_0,\ t = 0,\ldots\lambda)$

sont déterminés par les conditions $p_r^{(i)}/p_r^{(i)}$: $\mathcal{C}_m^{(i)}(h_s) = P(h_s)$, pour $s = i, i+1,\ldots i+m$ (si on part de $h_0 = H$, alors $i = 0$ évidemment), i.e. par

$$\mathcal{C}_{m,\delta}^{(i)} = P_\delta(h_s)$$

$$\delta = 1,2,\ldots n \tag{26}$$

(relations sur les composantes).

La valeur estimée de y(x) est alors: $y_c = C_m^{(i)}(0)$ que nous noterons C_m^i. Il est enfin possible de calculer les C_m^i par récurrence (triangle de Romberg). A chaque étape $(E_\alpha)_{\alpha=0,1,\ldots}$, nous calculons

$$\frac{1}{H^{2\alpha+2}} \prod_{s=0}^{\alpha} h_s^2 = \pi_\alpha = \frac{h_\alpha^2}{H^2} \pi_{\alpha-1} \tag{27}$$

et le processus est arrêté dès que $\pi_\alpha \leq \varepsilon$ (d'après les auteurs de la méthode).

Stabilisation des méthodes aux différences finies et autres améliorations

Toutes les méthodes conventionnelles à pas liés ou à pas séparés sont des intégrateurs exacts d'équations dans lesquelles les dérivées sont des polynômes de la variable indépendante (en négligeant les erreurs d'arrondis). Ainsi, un algorithme d'ordre N est défini habituellement comme intégrant exactement une équation différentielle dans laquelle le second membre est un polynôme de degré N. En pratique, un tel algorithme ne sera acceptable que lorsque le développement de Taylor de la fonction pourra être tronqué à l'ordre N avec une erreur négligeable (évaluée souvent à l'aide du terme de rang N+1). Divers auteurs (29,30), ont recherché des algorithmes qui intègrent exactement des fonctions non polynomiales. Ceci peut se faire, soit en modifiant dans ce but les coefficients A_p, B_p, A_p', B_p' d'un algorithme tel que celui décrit par les formules (13-15), soit en cherchant un changement de variable indépendante qui "régularise" un certain algorithme pour une famille donnée d'orbites. Illustrons successivement chacune des méthodes. Ecrivons les expressions (14) et (15) avec $P = 2$, $A_2 = A_2^* = 1$, $A_1 = A_1^* = -2$ et en se limitant à une fonction f ne dépendant que de x et t, sous la forme:

$$x_{n+1} = 2x_n - x_{n-1} + h^2 \sum_{p=j}^{N+j} \beta_p^{(j)} f_{n+1-p} \quad . \qquad (28)$$

L'indice supérieur (1) a été omis, étant sous-entendu que la valeur x_{n+1} obtenue par l'équation (14) doit être utilisée seulement dans le membre de droits de l'-équation (15); de plus, nous avons posé

$$\beta_p^{(1)} = B_p \quad , \quad \beta_p^{(0)} = B_p^* \quad ,$$

et l'équation (28) doit alors être utilisée successivement pour j = 1 et j = 0.

de l'équation (13), on déduit

$$x_{n+1} - 2x_n + x_{n-1} = \int_{t_n}^{t_{n+1}} dt \int_{t_n}^{t} f(x,t')dt'$$

$$+ \int_{t_n}^{t_{n-1}} dt \int_{t_n}^{t} f(x,t')dt' \quad ;$$

si l'on choisit $f(x,t) = \sum_{j=1}^{N+1} c_j g_j(t)$, on obtient

$$f_{n+1-p} = \sum_{j=1}^{N+1} c_j g_j(t_{n+1-p}) \quad , \quad p = 1,2,\ldots N+1 \quad , \qquad (29)$$

et donc, si: $G = \{g_j(t_p)\}$ est une matrice régulière, on a:

$$c_j = \sum_{p=1}^{N+1} G_{pj}^{-1} f_p$$

et enfin

$$x_{n+1} - 2x_n + x_{n-1} = h^2 \sum_{p=1}^{N+1} f_{n+1-p} \left\{ \sum_{j=1}^{N+1} G_{n+1-p,j}^{-1} \Gamma_j \right\} \qquad (30)$$

où

$$\Gamma_j = \frac{1}{h^2}\left\{\int_{t_n}^{t_{n+1}} dt \int_{t_n}^{t} g_j(t')dt' + \int_{t_n}^{t_{n-1}} dt \int_{t_n}^{t} g_j(t')dt'\right\}$$

L'équation (30) fournit, par comparaison avec (28), les coefficients cherchés $\beta_p^{(j)}$. Si donc nous voulons que (28) soit un intégrateur rigoureux (qui sera dit encore d'ordre N), des N+1 fonctions $f(x,t) = g_i(t)$, $(i = 1, \ldots N+1)$, il suffit de prendre les coefficients $\beta_p^{(j)}$ donnés par (30). Par exemple, les classiques algorithmes de Störmer et de Cowell correspondent à

$$g_i(t) = t^{i-1} \qquad (i = 1 \text{ à } N+1)$$

Le choix des $g_i(t)$ pour le mouvement Képlérien troublé est alors défini comme intégrant exactement jusqu'à l'ordre N le développement des coordonnées (ξ,η) du point dans son mouvement orbital non perturbé, en série de Fourier de l'anomalie moyenne M ou du temps t, avec M = nt. On peut alors prendre, par exemple (20)

$$g_i(t) = \left\{\begin{matrix} 1, \cos nt, \cos 2nt, \ldots \\ \sin nt, \sin 2nt, \ldots \end{matrix}\right\} \tag{31}$$

L'autre procédé, ainsi que nous l'avons dit, consiste à régulariser l'équation différentielle par un changement de variable descriptive du système.

Stiefel (31) et Moynot (24) utilisent la nouvelle variable σ définie, pour le problème perturbé des deux corps, par

$$\sigma = \int \frac{1}{r}\, dt \tag{32}$$

(transformation de Sundman particulière). En effet, on constate, dans le mouvement Képlérien, une erreur de troncature par les algorithmes usuels, d'autant plus importante que l'orbite est excentrique et que le point de départ est plus près du périgée, elle devient catastrophique pour des excentricités atteignant 0.8, surtout si on part du périgée. Il est alors naturel de chercher un moyen de restituer une orbite excentrique dans les mêmes conditions qu'une orbite circulaire. Or des formules telles que (14) ou (15) sont linéaires, donc insensibles à une affinité. Ce n'est donc pas l'ellipticité de l'orbite qui est gênante, mais la répartition des points calculés sur cette orbite: ils sont trop espacés au périgée, trop concentrés à l'apogée, et les techniques de changement de pas sont, comme nous l'avons signalé au début, souvent en retard par rapport à la nécessité numérique (moment de prendre la décision du changement de pas) et même conduisent à des instabilités. Pour rendre les points équidistants, à l'affinité près, le moyen le plus simple est de prendre comme variable indépendante, non plus le temps, mais l'anomalie excentrique, ou mieux, la quantité σ qui lui est proportionnelle.

L'équation du mouvement perturbé (13), où nous posons $x = \bar{r}$, est alors remplacée par le système

$$\begin{cases} \dfrac{dt}{d\sigma} = r \\[2ex] \dfrac{d^2\bar{r}}{d\sigma^2} = \dfrac{1}{r^2}\left(\bar{r}\cdot\dfrac{d\bar{r}}{d\sigma}\right)\dfrac{d\bar{r}}{d\sigma} + r^2\bar{f} \end{cases} \tag{33}$$

l'équation supplémentaire permettant de récupérer l'instant associé à une position donnée. Ayant ainsi fait choix de la variable indépendante, Moynot (24) améliore davantage la restitution de l'orbite de référence en modifiant l'algorithme utilisé (ici celui de Cowell), de façon à le mouvement Képlérien; le mouvement réel en étant très proche ne serait plus entaché que d'erreurs de troncature résiduelles dues uniquement à ces perturbations. On utilise donc une technique combinée de la régularisation (choix de σ) et de celle exposée précédemment (formules 28 à 31). Pour ce faire, on remarque que la solution Képlérienne s'explicite rigoureusement,

$$\bar{r} = \bar{a}_1 (\cos\nu_0\sigma - e) + \bar{a}_2 \sqrt{1-e^2} \sin\nu_0\sigma$$

$$t = t_0 + a\left\{\sigma - \frac{e}{\nu_0} \sin\nu_0\sigma\right\} \quad . \qquad (34)$$

Les deux vecteurs $\bar{a}_1$ et $\bar{a}_2$ sont orthogonaux et de module: le demi-grand axe a; e est l'excentricité, ν_0 est égal à $\sqrt{GM/a}$ et t_0 l'instant de passage au périgée. On rend donc l'algorithme de Cowell rigoureux pour toute solution dont chaque composante est de la forme: $x = x_0 + x_1 \cos\nu_0\sigma + x_2 \sin\nu_0\sigma$, admettant une partie constante et un terme sinusoidal de période égale à: $T_0 = 2\pi/\nu_0$ que l'on peut se fixer à l'avance.

L'équation du temps $dt/d\sigma = r$ pourra être intégrés rigoureusement dans les mêmes conditions si on retire l'effet séculaire; on n'intègrera donc pas l'équation du temps sous la forme ci-dessus, mais plutôt

$$\frac{dt'}{d\sigma} = r-a \qquad (35)$$

et on restituera le temps par $t = t' + a\sigma$. Pour les problèmes de géodésie spatiale et en ce qui concerne le potentiel terrestre, les erreurs de troncature du mouvement Képlérien étant résorbées, il est naturel d'examiner et de réduire les erreurs de troncature résiduelles, résultant de la plus importante des perturbations, à savoir celles dues à J_2 (32). On constate que, si l'on se limite aux orbites circulaires et si on fait varier l'inclinaison, les erreurs de troncature restent pratiquement nulles pour une orbite équatoriale et passent par un maximum pour une orbite polaire. En modifiant alors la variable indépendant, i.e. en remplaçant σ par

$$\sigma^* = \int \frac{1}{r^*} \, dt \quad , \qquad (36)$$

avec $r^* = \{x^2+y^2+(1+\varepsilon)z^2\}^{1/2}$, et en ajustant au démarrage la constante ε en fonction de J_2 et du demi-grand axe moyen a, on montre que l'on peut réduire d'à peu près un facteur 100 les erreurs de troncature!

Les améliorations dont il vient d'être question permettent, à précision égale et pour un temps de calcul inférieur ou égal, d'adopter des algorithmes d'intégration d'ordre moins élevé.

Une autre méthode, appliquée à l'algorithme de Bulirsch et Stoer, consiste à accélérer la convergence de la boucle prédiction-correction, en remarquant que le formalisme de prédiction est linéaire par morceaux. Le prédictor est donc un intégrateur rigoureux du mouvement Képlérien pour les équations du mouvement, écrites en a, e, I, Ω, ω, M (demi-grand axe, excentricité, inclinaison, longitude du noeud, argument du périgée, anomalie moyenne), c'est-à-dire pour les équations de Gauss (ou de Lagrange) de la mécanique céleste (7).

Quelques comparaisons

Il n'y a bien sûr pas de méthode d'intégration miracle. Pour comparer l'efficacité de deux méthodes eu égard à des contraintes de temps de calcul, de précision, de difficulté de mise en oeuvre etc..., il faut définir plusieurs classes de problèmes.

Mouvement Képlérien

Nous donnons quelques indices de comparaison entre les méthodes de Runge-Kutta (ordre 4) et de Taylor d'une part, d'après Roy et al. (6), et entre la méthode de Cowell classique d'ordre 8 et celle utilisant la régularisation et la stabilisation décrites entre les formules (32) à (35), d'après Moynot (24), voir tables 5, 6, 7.

La méthode de Cowell n'est donc supérieure que pour des précisions très grandes, dans ce cas.

Mouvements à N - corps - Amas globulaires

Quelques simulation effectuées par Black (9) et utilisant un algorithme récursif par séries de Taylor, ont montré la possibilité de formation de systèmes binaires, triples etc...et des quasi collisions, où le pas était parfois divisé par 10^{15}! Le non emploi, à ce stade, de méthodes de régularisation fait douter de la validité de la solution. Cependant, une intégration numérique en sens inverse a chaque fois a montré la formation de ces mêmes systèmes (en sens inverse), bien que les conditions initiales ne soient retrouvées qu'avec une précision relative de 10^{-2} à 10^{-3}.

Nous avons par ailleurs comparé les résultats donnés par la méthode mixte d'Adams-Cowell mise en oeuvre par Oesterwinter (25), et celle de Bulirsch et Stoer utilisée soit en coordonnées rectangulaires pour le mouvement des 9 planètes du système solaire, soit pour les èquations de Lagrange dans le cas de la Lune, et avons trouvé une remarquable stabilité de la méthode d'extrapolation de Bulirsch et Stoer vis à vis d'une méthode classique à pas liés d'ordre 14 (que nous avons pris comme référence).

L'efficacité de la méthode de Bulirsch et Stoer a par ailleurs été comparée à celle de la méthode de Cowell régularisée d'ordre 8 pour le mouvement d'un satellite artificiel de la Terre soumis à tout le potentiel terrestre et pour l'intégration des équations aux variations relatives aux conditions initiales et aux harmoniques sphériques du potentiel. La méthode de Bulirsch et Stoer semble plus précise, bien qu'un peu plus coûteuse en temps calcul.

D'une manière générale, une étude très complète de nombreux algorithmes d'intégration numérique confrontés à des problèmes très divers (33), a montré irréfutablement que la méthode de Bulirsch et Stoer est la plus efficace tant que le nombre d'opérations élémentaires nécessaires au calcul de la fonction, second membre du système à integrer, n'est pas trop élevé (inférieur en moyenne à 50 opérations arithmétiques par composante). Sinon l'avantage est à des formules du type Adams, ou Cowell régularisé....

Calcul des dérivées partielles

Le problème consiste à calculer, pour un système différentiel mis sous l'une des formes suivantes

$$\dot{x} = X(t,x,\varepsilon) \tag{37}$$

$$\ddot{x} = F(t,x,\dot{x},\varepsilon) \tag{38}$$

avec, pour t_0: x_0, $\dot{x}_0$ donnés (ou calculés), les quantités

$$\frac{\partial x}{\partial x_0}, \frac{\partial \dot{x}}{\partial x_0}, \frac{\partial x}{\partial \dot{x}_0}, \frac{\partial \dot{x}}{\partial \dot{x}_0}, \frac{\partial x}{\partial \varepsilon}, \frac{\partial \dot{x}}{\partial \varepsilon}, \varepsilon$$

étant un vecteur de paramètres physiques dont dépend la trajectoire. $\hat{x}(t,t_0,x_0,\dot{x}_0,\varepsilon)$ sera la solution du système considéré.

Méthode par différences finies

Quel que soit le paramètre q par rapport auquel on veut calculer une dérivée partielle, que ce soit une composante de x_0 ou $\dot{x}_0$, ou l'un des ε, il est toujours possible d'approximer $\partial x/\partial q$ par

$$\frac{\partial x}{\partial q} \simeq D_q x = \frac{\hat{x}(t,\ldots,q+\delta q,\ldots) - \hat{x}(t,\ldots,q,\ldots)}{\delta q} \tag{39}$$

à condition que la solution $\hat{x}$ soit de classe C^1 au moins dans le domaine étudié. La précision dépend évidemment de l'allure de la fonction, de sa sensibilité à chaque paramètre q, et finalement du choix de δq. Cette méthode est très lourde car elle nécessite l'intégration d'un système du type (37) ou (38) supplémentaire pour chaque paramètre q, avec des conditions initiales donc différentes. C'est pourquoi elle n'est en général utilisée que pour les dérivées partielles par rapport aux éléments initiaux $(x_0,\dot{x}_0)$ et encore à l'aide d'une fonction F simplifiée. Dans le cas de la détermination d'orbite d'un satellite artificiel proche de la Terre, par exemple, par correction différentielle itérative, on se contente pour les dérivées, au lieu de toute la fonction perturbatrice, d'une fonction F ne prenant en compte que les quelques premiers harmoniques zonaux du potentiel gravitationnel (plus éventuellement le frottement et la pression de radiation). Ceci est en général suffisant, bien que ralentissant la convergence du processus d'approximations successives.

Les équations aux variations

En dérivant le système (37) par rapport à un paramètre quelconque q, on obtient aisément

$$\frac{d}{dt}\left(\frac{\partial x}{\partial q}\right) = \frac{\partial X}{\partial \hat{x}} \frac{\partial \hat{x}}{\partial q} + \frac{\partial X}{\partial q} \quad , \tag{40}$$

où l'on a remplacé x par la solution intégrée de (37), $\hat{x}$. Si q est un paramètre tel que x_0^i ou $\dot{x}_0^i$, cela se réduit à

$$\frac{d}{dt}\left(\frac{\partial \hat{x}}{\partial q}\right) = \frac{\partial X}{\partial \hat{x}} \frac{\partial \hat{x}}{\partial q} \quad . \tag{41}$$

Dans le premier cas, les conditions initiales sont évidemment nulles, dans le second, ce sont

$$\frac{D(x,\dot{x})}{D(x_0,\dot{x}_0)} = \text{matrice unité de rang 6} \quad .$$

Dans le cas du système (40), les équations aux variations prennent la forme:

$$\frac{d^2}{dt^2}\left(\frac{\partial \hat{x}}{\partial q}\right) = \frac{\partial F}{\partial \hat{x}} \frac{\partial x}{\partial q} + \frac{\partial F}{\partial \hat{\dot{x}}} \frac{d}{dt}\left(\frac{\partial \hat{x}}{\partial q}\right) + \frac{\partial F}{\partial q} \quad . \tag{42}$$

Sous une forme ou sous une autre, elles constituent donc un système d'équations supplémentaires qui peut-être intégré simultanément aux équations du mouvement. Ceci a pour avantage essentiel de n'adopter qu'un algorithme d'intégration numérique pour le problème traité, de calculer, à chaque pas, des quantités qui servent à la fois aux deux systèmes et donc d'économiser beaucoup de temps (surtout si les seconds membres nécessitent beaucoup) plus d'opérations élémentaires que l'algorithme). L'inconvénient majeur réside dans le fait que les dérivées partielles varient souvent bien plus lentement que les fonctions x et $\dot{x}$ elles-mêmes, et donc qu'il est plus astucieux d'employer des algorithmes d'ordre moins élevé, ou à plus grand pas, et donc plus rapides à précision égale. Mais l'utilisation simultanée de deux algorithmes différents est souvent mal aisée, et l'on préfère souvent traiter le problème à part, en se servant d'une tabulation de x et $\dot{x}$ obtenue lors d'une intégration préalable. Cependant, quelle que soit la difficulté de mise en oeuvre, les équations aux variations demeurent la base de tout calcul de dérivées partielles et leur intégration est une référence précieuse pour la comparaison à d'autres méthodes ou variantes.

La méthode de Bulirsch et Stoer est bien adaptée à l'intégration numérique simultanée des équations du mouvement d'un satellite artificiel proche de la Terre, il est possible de prendre des pas "globaux" H de l'ordre du quart de révolution, le pas de calcul réel descendant, lui, jusqu'à une minute, une demi-minute ou moins, suivant les exigences de précision (puisque, ainsi qu'on l'a vu, l'ordre de la méthode, qui est le nombre d'étapes effectuées avec un pas h_s, est variable et dépend en chaque point de la rapidité de convergence de l'extrapolation à zéro des fractions rationnelles). Il est, de plus, possible de pondérer le critère de convergence de façon à converger essentiellement sur l'orbite et non pas (ou très peu apparemment), sur les dérivées partielles (...qui s'avèrent finalement être justes). Ainsi, le temps d'intégration est à peine multiplié par 3 dans le cas d'équations aux variations pour les 6 conditions initiales (position et vitesse) et pour 120 coefficients du potentiel, par rapport au temps d'intégration du système (37) seul.

Une autre manière de procéder est de remarquer que le système (41) est de la forme

$$\dot{y} = Uy \quad , \qquad (43)$$

avec

$$U = \frac{\partial X}{\partial \hat{x}} = U(t) \quad ,$$

$$y = \frac{\partial \hat{x}}{\partial q} \quad , \quad y_0 = \left(\frac{\partial x}{\partial q}\right)_{t_0} \quad .$$

Définissant alors le noyau résolvant de (43), ou matrizant, par

$$\Omega = \Omega(t,t_0;U) = I + \int_{t_0}^{t} U(\tau)d\tau$$

$$+ \int_{t_0}^{t} U(\sigma)\int_{t_0}^{\sigma} U(\tau)d\tau d\sigma + \ldots \quad , \qquad (44)$$

i.e. par l'intégrale de la série de Neumann de U, nous avons les propriétés suivantes

$$\Omega(t_0,t_0;U) = I$$

$$\frac{d\Omega}{dt} = U\Omega$$

$y = \Omega y_0$ est la solution de (43) . (45)

De plus, si maintenant on cherche à intégrer le système

$$\dot{y} = Uy + z(t)$$

(qui est la forme du système (40)), alors sa solution est (34)

$$y = \Omega y_0 + \Omega\int_{t_0}^{t} \Omega^{-1} z(\tau)d\tau \quad . \qquad (46)$$

Enfin, si le système (43) peut-être écrit sous la forme:

$$\frac{dy}{dt} = \dot{y} = (U+V)y \qquad (47)$$

avec $V = V(t)$. Alors, si l'on pose

$$y = \Omega(t,t_0,U)x \quad (\text{donc } y_0 = x_0) \quad . \tag{48}$$

x vérifie le système

$$\dot{x} = \Omega^{-1} V\Omega x \quad . \tag{49}$$

On démontre également (35) que

$$\Omega(t,t_0,U) = \Omega(t,t_1,U)\Omega(t_1,t_0,U)$$

et

$$\Omega(t,t_0,U) = \left[\Omega(t_0,t,U)\right]^{-1} \quad . \tag{50}$$

Ces dernières relations montrent que Ω est déterminé dès qu'il est calculé pour un intervalle quelconque (t_0, t), par exemple sur quelques pas d'intégration. C'est ce qui permet de le trouver rapidement en limitant la procédure d'intégration numérique de (45) à quelques orbites en pratique.

Cependant, cette méthode utilisant la continuation "en chaine" par produits de matrice, se dégrade en général rapidement en précision, du fait d'une propagation catastrophique des erreurs d'arrondis.

On utilise alors la variante, ci-après, qui est une application de la méthode de Peano-Baker, telle qu'elle a été mise en oeuvre par H. Ditto (36) et B. Moynot (24).

Considérons les équations (38) pour le mouvement Képlérien perturbé, sous la forme

$$\ddot{\bar{r}} = -\frac{\mu}{r^3}\bar{r} + \bar{g}(\bar{r},\dot{\bar{r}},\bar{\varepsilon},t) \tag{51}$$

avec les conditions initiales $\bar{r}_0$, $\dot{\bar{r}}_0$; μ est une constante positive (constante gravitationnelle), $\bar{g}$ est le vecteur des forces non Képlériennes, $\bar{\varepsilon}$ le vecteur des paramètres inconnus du modèle de forces. Les équations (42) s'écrivent, sous forme canonique:

$$\frac{d}{dt}\left(\frac{\partial \bar{r}}{\partial \bar{r}_0}\right) = \frac{\partial \dot{\bar{r}}}{\partial \bar{r}_0}$$

$$\frac{d}{dt}\left(\frac{\partial \dot{\bar{r}}}{\partial \bar{r}_0}\right) = G \frac{\partial \bar{r}}{\partial \bar{r}_0} + \frac{\partial \bar{g}}{\partial \bar{r}} \frac{\partial \bar{r}}{\partial \bar{r}_0} + \frac{\partial \bar{g}}{\partial \dot{\bar{r}}} \frac{\partial \dot{\bar{r}}}{\partial r_0} \qquad (52)$$

avec

$$G = \frac{\mu}{r^5} (3\ \bar{r}\ \bar{r}^T - \bar{r} \times \bar{r}\ I) \qquad (53)$$

et l'on a des équations similaires pour les dérivées par rapport à $\dot{\bar{r}}_0$ et $\bar{\varepsilon}$.

Nous avons posé

$$\frac{\partial \bar{r}}{\partial \bar{r}_0} = \frac{D(x,y,z)}{D(x_0,y_0,z_0)} \quad , \quad \frac{\partial \bar{g}}{\partial \bar{r}} = \frac{D(g_x g_y g_z)}{D(x,y,z)} \quad , \text{ etc...}$$

$$\bar{r}\ \bar{r}^T = \begin{pmatrix} x \\ y \\ z \end{pmatrix} (x,y,z) = \begin{pmatrix} x^2 & xy & xz \\ yx & y^2 & yz \\ zx & zy & z^2 \end{pmatrix}$$

$$I = \text{matrice unité } 3 \times 3 \quad .$$

L'ensemble des équations aux variations peut être résumé par

$$\frac{d}{dt}\begin{bmatrix} \frac{\partial \bar{r}}{\partial \bar{r}_0} & \frac{\partial \bar{r}}{\partial \dot{\bar{r}}_0} & \frac{\partial r}{\partial \bar{\varepsilon}} \\ \frac{\partial \dot{\bar{r}}}{\partial \bar{r}_0} & \frac{\partial \dot{\bar{r}}}{\partial \dot{\bar{r}}_0} & \frac{\partial \dot{\bar{r}}}{\partial \bar{\varepsilon}} \end{bmatrix} = \left\{ \begin{bmatrix} 0 & I \\ G & 0 \end{bmatrix} + \begin{bmatrix} 0 & 0 \\ \frac{\partial \bar{g}}{\partial \bar{r}} & \frac{\partial \bar{g}}{\partial \dot{\bar{r}}} \end{bmatrix} \right\}$$

$$\times \begin{bmatrix} \frac{\partial \bar{r}}{\partial \bar{r}_0} & \frac{\partial \bar{r}}{\partial \dot{\bar{r}}_0} & \frac{\partial \bar{r}}{\partial \bar{\varepsilon}} \\ \frac{\partial \dot{\bar{r}}}{\partial \bar{r}_0} & \frac{\partial \dot{\bar{r}}}{\partial \dot{\bar{r}}_0} & \frac{\partial \dot{\bar{r}}}{\partial \bar{\varepsilon}} \end{bmatrix} + \begin{bmatrix} 0 & 0 & 0 \\ 0 & 0 & \frac{\partial \bar{g}}{\partial \bar{\varepsilon}} \end{bmatrix} \quad . \qquad (54)$$

On recherche alors le matrizant du mouvement Képlérien, c'est-à-dire

$$\Omega(t,t_0,C) = \begin{bmatrix} \frac{\partial \bar{r}}{\partial \bar{r}_0} & \frac{\partial \bar{r}}{\partial \dot{\bar{r}}_0} \\ \frac{\partial \dot{\bar{r}}}{\partial \bar{r}_0} & \frac{\partial \dot{\bar{r}}}{\partial \dot{\bar{r}}_0} \end{bmatrix} ,$$

vérifiant

$$\frac{d\Omega(t,t_0,C)}{dt} = \begin{bmatrix} 0 & I \\ C & 0 \end{bmatrix} \Omega(t,t_0,C) \quad , \qquad (55)$$

pour

$$\ddot{\bar{r}} = -\mu\bar{r}/r^3 \qquad (56)$$

$$C = \frac{\mu}{r^5}\,(3\;\bar{r}\;\bar{r}^T - \bar{r}\cdot\bar{r}\;I) \quad . \qquad (57)$$

On reporte évidemment la solution de (56) dans (57).
Posant

$$\Omega(t,t_0,C) = \begin{bmatrix} L(t,t_0) & M(t,t_0) \\ P(t,t_0) & Q(t,t_0) \end{bmatrix} \quad . \qquad (58)$$

Nous avons

$$\left[\Omega(t,t_0,C)\right]^{-1} = \Omega(t_0,t,C)$$

$$= \begin{bmatrix} Q^T(t,t_0) & -M^T(t,t_0) \\ -P^T(t,t_0) & L^T(t,t_0) \end{bmatrix} \quad . \tag{59}$$

Le matrizant $\Omega(t,t_0,C)$ peut évidemment être calculé à l'aide du formalisme analytique du problème des deux corps.

Posons maintenant:

$$U = \begin{bmatrix} 0 & I \\ C & 0 \end{bmatrix} \quad \text{et} \quad \Omega(t,t_0,C) \equiv \Omega(t,t_0,U)$$

$$V = \begin{bmatrix} 0 & 0 \\ G-C & 0 \end{bmatrix} + \begin{bmatrix} 0 & 0 \\ \frac{\partial \bar{g}}{\partial \bar{r}} & \frac{\partial \bar{g}}{\partial \bar{r}} \end{bmatrix}$$

$$y = \text{solution cherchée} = \begin{bmatrix} \frac{\partial \bar{r}}{\partial \bar{r}_0} & \frac{\partial \bar{r}}{\partial \dot{\bar{r}}_0} \\ \frac{\partial \dot{\bar{r}}}{\partial \bar{r}_0} & \frac{\partial \dot{\bar{r}}}{\partial \dot{\bar{r}}_0} \end{bmatrix}$$

$$x = \begin{bmatrix} X & Y \\ Z & W \end{bmatrix}$$

$$y = \Omega(t,t_0,C)x \quad . \tag{60}$$

Nous avons d'après (49)

$$\frac{d}{dt}\begin{bmatrix} X & Y \\ Z & W \end{bmatrix} = \Omega(t,t_0,C)^{-1}$$

$$\times \left\{ \begin{bmatrix} 0 & 0 \\ G-C & 0 \end{bmatrix} + \begin{bmatrix} 0 & 0 \\ \frac{\partial \bar{g}}{\partial r} & \frac{\partial \bar{g}}{\partial \dot{\bar{r}}} \end{bmatrix} \right\} \Omega(t,t_0,C) \begin{bmatrix} X & Y \\ Z & W \end{bmatrix} \quad . \tag{61}$$

Les équations (61) fournissent donc par intégration une matrice de correction au matrizant Képlérien, et donc la matrice des dérivées partielles du mouvement perturbé. L'avantage d'utiliser ces équations réside dans le fait que l'on peut employer à la fois des pas plus grands et des algorithmes d'ordre nettement inférieur à celui de l'intégration des équations du mouvement.

La matrice des dérivées partielles cherchée est donc

$$\Omega(t,t_0,G) = \Omega(t,t_0,C) \begin{bmatrix} X & Y \\ Z & W \end{bmatrix} \quad . \tag{62}$$

Quant aux dérivées par rapport aux paramètres $\bar{\varepsilon}$, elles vérifient

$$\frac{d}{dt}\begin{bmatrix} \frac{\partial \bar{r}}{\partial \bar{\varepsilon}} \\ \frac{\partial \dot{\bar{r}}}{\partial \bar{\varepsilon}} \end{bmatrix} = \left\{ \begin{bmatrix} 0 & I \\ G & 0 \end{bmatrix} + \begin{bmatrix} 0 & 0 \\ \frac{\partial \bar{g}}{\partial \bar{r}} & \frac{\partial \bar{g}}{\partial \dot{\bar{r}}} \end{bmatrix} \right\}$$

$$\times \begin{bmatrix} \frac{\partial \bar{r}}{\partial \varepsilon} \\ \frac{\partial \dot{\bar{r}}}{\partial \bar{\varepsilon}} \end{bmatrix} + \begin{bmatrix} 0 \\ \frac{\partial \bar{g}}{\partial \bar{\varepsilon}} \end{bmatrix} \quad . \tag{63}$$

D'après (46) la solution est, puisque l'on impose en général

$$\frac{\partial \bar{r}}{\partial \bar{\varepsilon}} = \frac{\partial \dot{\bar{r}}}{\partial \bar{\varepsilon}} = 0$$

à l'instant initial:

$$\begin{bmatrix} \dfrac{\partial \bar{r}}{\partial \bar{\varepsilon}} \\ \dfrac{\partial \dot{\bar{r}}}{\partial \bar{\varepsilon}} \end{bmatrix} = \Omega(t,t_0,G) \int_{t_0}^{t} \left[\Omega(\tau,t_0,G)\right]^{-1} \begin{bmatrix} 0 \\ \dfrac{\partial \bar{g}}{\partial \bar{\varepsilon}}(\tau) \end{bmatrix} d\tau \quad . \qquad (64)$$

Remarque: Nous avons utilisé cette méthode avec l'algorithme de Cowell d'ordre 8, régularisé ainsi qu'il a été dit au précédent paragraphe. Le choix de la variable σ pour l'orbite de référence transforme évidemment les équations (61) et (64); par exemple,

$$\begin{bmatrix} \dfrac{\partial \bar{r}}{\partial \bar{\varepsilon}} \\ \dfrac{\partial \dot{\bar{r}}}{\partial \bar{\varepsilon}} \end{bmatrix} = \Omega(t,t_0,G) \int_{\sigma_0}^{\sigma} \Omega^{-4}(t(\sigma'),t\ ,G)$$

$$\times \begin{bmatrix} 0 \\ \dfrac{\partial \bar{g}}{\partial \bar{\varepsilon}} \left[t(\sigma')\right] \end{bmatrix} r(\sigma')\,d\sigma' \quad . \qquad (65)$$

Les équations (61) et (65) sont intégrées par la méthode de Cowell adaptée aux systèmes du 1er ordre. Le calcul des dérivées par rapport aux ε, à σ donné, peut se faire du premier coup: la séquence d'intégration sur un pas se réduit donc à la séquence de prédiction, d'où un gain de temps très considérable par rapport à l'algorithme complet.

Méthodes Rapides de Prediction (mouvement Képlérien perturbé)

Depuis quelques années, les techniques d'intégration numérique de systèmes aux éléments moyens, ou consistant à moyenner, en général par quadrature, certains effets perturbateurs de période inférieure à une période donnée (celle de l'orbite moyenne dans le problème des deux corps), connaissent une grande popularité, car elles permettent d'analyser les perturbations d'une trajectoire, souvent avec assez de précision, pendant des intervalles de temps formidablement longs (plusieurs années) sans avoir recours aux coûteux algorithmes classiques d'ordre élevé qui ont été décrits au premier paragraphe. Les applications sont très nombreuses: études de missions planétaires, fenêtres de tir, durées de vie...en général toutes les études paramétriques pour lesquelles on ne dispose malheureusement pas de théorie analytique.

Le principe en est le suivant: soit E un élément orbital quelconque, $\dot{E}$ sa dérivée par rapport au temps. Considérons la quantité, à chaque instant t:

$$\bar{\dot{E}} = \frac{1}{\tau} \int_{t-\tau/2}^{t+\tau/2} \dot{E}(t)\,dt \qquad (66)$$

où τ est la période orbitale moyenne définie par

$$\tau = \frac{2\pi}{\langle n\rangle} = 2\pi \left(\frac{\langle a\rangle^3}{\mu}\right)^{1/2} \quad . \qquad (67)$$

En prenant l'anomalie vraie, v, comme variable indépendante, l'équation (66) devient

$$\bar{\dot{E}} = \sqrt{\frac{\mu}{a^3}}\ \frac{a^2(1-e^2)^2}{2\pi\sqrt{\mu a(1-e^2)}} \int_{v(t-\tau/2)}^{v(t+\tau/2)} (1+e\ \cos\ v)^{-2}\ \dot{E}(v)\,dv\ . \tag{68}$$

On considère alors, et c'est la base de la méthode que les éléments orbitaux sont constants pendant le temps d'intégration et peuvent donc être sortis du signe somme; on suppose donc le mouvement non perturbé pendant l'intervalle de temps de l'intégration. Par exemple, en utilisant la formulation des équations de Gauss, nous avons pour les variables i (inclinaison), Ω (longitude du noeud), $\psi = \omega+v$ (longitude vraie dans l'orbite, comptée à partir du périgée), d'après Uphoff (37):

$$\overline{\frac{di}{dt}} = np^2(2\pi\mu)^{-1} \int_{v_1}^{v_2} W(v)\cos\psi(1+e\ \cos\ v)^{-3}dv$$

$$\bar{\dot{\Omega}} = \overline{\frac{d\Omega}{dt}} = np^2(2\pi\mu) \int_{v_1}^{v_2} W(v)\sin\psi\Big/\left[\sin\ i(1+e\cos v)^3\right]dv$$

$$\bar{\dot{\psi}} = \overline{\frac{d\psi}{dt}} = n-np^2(2\pi\mu)^{-1}$$

$$\times\ \mathrm{cotg}\ i \int_{v_1}^{v_2} W(v)\sin\psi(1+e\ \cos\ v)^{-3}dv\ . \tag{69}$$

n est le moyen mouvement, $p = a(1-e^2)$, W est la composante des forces perturbatrices normale au plan de l'orbite osculatrice. Pour les autres éléments, nous considérerons $\bar{\dot{p}} = \overline{\frac{dp}{dt}}$, $\bar{\dot{h}} = \overline{\frac{d}{dt}(e\ \sin\omega)}$ et $\bar{\dot{k}} = \overline{\frac{d}{dt}(e\ \cos\omega)}$ afin de ne pas exclure les orbites d'excentricité nulle. R et S étant les deux autres composantes des forces perturbatrices, situées dans le plan de l'orbite (composantes radiale et perpendiculaire, respectivement), nous avons

$$\bar{\dot{p}} = np^3(\mu\pi)^{-1} \int_{v_1}^{v_2} (1+e \cos v)^{-3} S(v)dv \quad ,$$

$$\bar{\dot{h}} = np^2(2\pi\mu)^{-1} \int_{v_1}^{v_2} \Big\{-R(v)\cos\psi(1+e \cos v)^{-2}$$

$$- k \sin\psi \text{cotg } i\ W(v)\cdot(1+e \cos v)^{-3}$$

$$+ \Big[(2+e \cos v)\sin\psi+h\Big](1+e \cos v)^{-3} S(v)\Big\} dv \quad ,$$

$$\bar{\dot{k}} = np^2(2\pi\mu)^{-1} \int_{v_1}^{v_2} \Big\{R(v)\sin\psi(1+e \cos v)^{-2} + h \sin\psi$$

$$\times \text{cotg } i\ W(v)\cdot(1+e \cos v)^{-3}$$

$$+ \Big[(2+e \cos v)\cos\psi+k\Big](1+e \cos v)^{-3} S(v)\Big\} dv \quad . \tag{70}$$

Les équations (69) et (70) où $v_1 = v(t-\tau/2)$ et $v_2 = v(t+\tau/2)$, peuvent alors être utilisées avec une bonne méthode de quadrature (Gauss par exemple) et une méthode d'intégration numérique stable, à très grand pas. Divers auteurs ont utilisé cette méthode avec différentes variantes et des aménagements personnels (voir par exemple Balmino (38), Roth (39)). On peut par exemple, à l'aide de l'algorithme de Bulirsch et Stoer appliqué aux équations (69) et (70), utiliser sans inconvénient des pas de 10 à 100 révolutions pour un satellite de la Terre, suivant la précision désirée. Il faut évidemment avoir fait choix auparavant de la période avec assez de précision, par exemple prendre la période du mouvement Képlérien troublé par l'aplatissement terrestre.

Cette méthode sert particulièrement bien les études des phénomènes à longue période, mais la même technique peut aussi s'appliquer, ainsi que l'a montré Roth (39), aux termes à courte période en remplaçant l'intégration entre $v_1(t-\tau/2)$ et $v_2(t+\tau/2)$ par celle entre $v_1(t-\theta)$ et $v_2(t+\theta)$, où θ est arbitrairement petit.

Table 1. Coefficients de Runge-Kutta - Ordre 4 (Kutta)

0	-			
1/2	1/2	-		
1/2	0	1/2	-	
1	0	0	1	-
	1/6	1/3	1/3	1/6

Table 2. Coefficients de Runge-Kutta - Ordre 6 (17)

0	-						
$\frac{5-\sqrt{5}}{10}$	$\frac{5-\sqrt{5}}{10}$	-					
$\frac{5+\sqrt{5}}{10}$	$\frac{5}{\sqrt{10}}$	$\frac{5+2\sqrt{5}}{10}$	-				
$\frac{5-\sqrt{5}}{10}$	$\frac{-15+7\sqrt{5}}{20}$	$\frac{-1+\sqrt{5}}{4}$	$\frac{15-7\sqrt{5}}{10}$	-			
$\frac{5+\sqrt{5}}{10}$	$\frac{5-\sqrt{5}}{60}$	0	$\frac{1}{6}$	$\frac{15+7\sqrt{5}}{60}$	-		
$\frac{5-\sqrt{5}}{10}$	$\frac{5+\sqrt{5}}{60}$	0	$\frac{9-5\sqrt{5}}{12}$	$\frac{1}{6}$	$\frac{-5+3\sqrt{5}}{10}$	-	
1	$\frac{1}{6}$	0	$\frac{-55+25\sqrt{5}}{12}$	$\frac{-25-7\sqrt{5}}{12}$	$\frac{5-2\sqrt{5}}{10}$	$\frac{5+\sqrt{5}}{2}$	-
	$\frac{1}{2}$	0	0	$\frac{5}{12}$	$\frac{5}{12}$	$\frac{5}{12}$	$\frac{1}{12}$

Table 3. Coefficients de Runge-Kutta - Ordre 8 (18)

0	-											
1/9	$\frac{1}{9}$	-										
1/6	$\frac{1}{24}$	$\frac{1}{8}$	-									
1/4	$\frac{1}{16}$	0	$\frac{3}{16}$	-								
1/10	$\frac{29}{500}$	0	$\frac{39}{500}$	$-\frac{3}{125}$	-							
1/6	$\frac{33}{972}$	0	0	$\frac{4}{972}$	$\frac{125}{972}$	-						
1/2	$-\frac{21}{36}$	0	0	$\frac{76}{36}$	$\frac{125}{36}$	$-\frac{162}{36}$	-					
2/3	$-\frac{30}{243}$	0	0	$-\frac{32}{243}$	$\frac{125}{243}$	0	$\frac{99}{243}$	-				
1/3	$\frac{1175}{324}$	0	0	$-\frac{3456}{324}$	$-\frac{6250}{324}$	$\frac{8424}{324}$	$\frac{242}{324}$	$-\frac{27}{324}$	-			
5/6	$\frac{293}{324}$	0	0	$-\frac{852}{324}$	$\frac{1375}{324}$	$\frac{1836}{324}$	$-\frac{118}{324}$	$\frac{162}{324}$	1	-		
5/6	$\frac{1303}{1620}$	0	0	$-\frac{4260}{1620}$	$-\frac{6875}{1620}$	$\frac{9990}{1620}$	$\frac{1030}{1620}$	0	0	$\frac{1}{10}$	-	
1	$-\frac{8595}{4428}$	0	0	$-\frac{30720}{4428}$	$\frac{48750}{4428}$	$-\frac{66096}{4428}$	$\frac{378}{4428}$	$-\frac{729}{4428}$	$-\frac{1944}{4428}$	$-\frac{1296}{4428}$	$\frac{3240}{4428}$	-
	$\frac{41}{840}$	0	0	0	0	$\frac{216}{840}$	$\frac{272}{840}$	$\frac{27}{840}$	$\frac{27}{840}$	$\frac{36}{840}$	$\frac{180}{840}$	$\frac{41}{840}$

Table 4. Coefficients de Adams et Cowell (d'après Oesterwinter et al. (25)).

j	a_j	c_j
0	+32 011 868 528 640 000	+32 011 868 528 640 000
1	-16 005 934 264 320 000	-32 011 868 528 640 000
2	-2 667 655 710 720 000	+2 667 655 710 720 000
3	-1 333 827 855 360 000	000
4	- 844 757 641 728 000	- 133 382 785 536 000
5	- 600 222 534 912 000	- 133 382 785 536 000
6	- 456 783 110 784 000	- 116 974 585 728 000
7	- 363 891 528 000 000	- 100 566 385 920 000
8	- 299 520 219 398 400	-86 707 632 211 200
9	- 252 655 401 398 400	-75 398 324 601 600
10	- 217 227 737 563 200	-66 193 573 118 400
11	- 189 640 115 028 000	-58 648 487 788 800
12	- 167 636 336 098 320	-52 401 453 198 480
13	- 149 735 464 049 160	-47 174 128 491 600
14	- 134 928 496 929 540	-42 755 108 505 900
15	- 122 506 205 369 730	-38 983 584 907 800
16	- 111 956 703 448 001	-35 736 323 456 205

D = 32 011 868 528 640 000.

Tous les nombres doivent être divisés par le dénominateur commun D.

Table 5. Erreurs relatives après 50 orbites, pour les méthodes de Runge-Kutta (d'ordre 4), et de la série de Taylor (n=15), pour $e_\ell = 10^{-11}$.

	Erreurs sur les composantes $\times 10^{10}$	Erreur sur le demi-grand axe $\times 10^{10}$	Erreur moyenne sur les composantes du moment cinétique	Nombre de pas utilisés
e = 0. (Rk4	2.8	15	2.7	37314
e = 0. (T.S.	3.6	2	0.5	684
e = 0.95 (Rk4	1442.0	11671	176.0	84016
e = 0.95 (T.S.	80.0	1921	1.6	4905

Table 6. Erreurs après 5 orbites, pour la méthode de Cowell classique (ordre 8) et la même méthode régularisée (demi-grand axe = 8600 km)

Point de Départ	e	Δa (mètres)	ΔM (radians)	Δa (mètres)	ΔM radians
Périgée	0.8	$>2.10^{7}$	plusieurs fois 2π (!)		$-0.82.10^{-11}$
	0.6	-2.15	$-0.79.10^{-3}$	$0.61.10^{-7}$	$-1.79\ 10^{-12}$
	0.4	$-2.12\ 10^{-4}$	$-0.50.10^{-6}$	$1.06.10^{-7}$	$-0.52\ 10^{-12}$
	0.2	$-1.05\ 10^{-6}$	$-0.90.10^{-9}$	$1.34.10^{-7}$	$-1.76\ 10^{-13}$
	0.	$-0.33\ 10^{-8}$	$-0.52.10^{-13}$	$1.33.10^{-7}$	$-0.61\ 10^{-13}$
Apogée	0.2	$-1.05\ 10^{-6}$	$1.08\ 10^{-13}$	$1.27.10^{-7}$	$-1.89.10^{-14}$
	0.4	$-2.09\ 10^{-4}$	$0.91\ 10^{-10}$	$1.21.10^{-7}$	$-0.37.10^{-14}$
	0.6	$-0.65\ 10^{-2}$	$-2.41\ 10^{-9}$	$1.16.10^{-7}$	$-1.32.10^{-15}$
	0.8	-598	$1.88.10^{-5}$	$1.15.10^{-7}$	$0.50.10^{-14}$
	Cowell classique			Cowell régularisé	

La conclusion s'impose d'elle-même!

Table 7. Efficacité relative de la méthode de Cowell et de celle de Taylor (intégration des équations d'Encke 1.18) pour différentes précisions demandées et pour divers $\mu = m/M$

	μ	$\varepsilon = 10^{-5}$	$\varepsilon = 10^{-11}$	$\varepsilon = 10^{-15}$
$e = 0.1$	10^{-6}	0.7	0.5	1.2
	10^{-5}	0.2	0.3	2.2
	10^{-4}	0.1	0.5	1.8
	10^{-3}	0.5	0.2	5.1
$e = 0.9$	10^{-6}	0.5	0.2	3.6
	10^{-5}	0.7	0.2	5.6
	10^{-4}	0.3	0.4	3.4
	10^{-3}	0.5	1.9	6.1

N.B. - Les nombres de la table sont approximativement proportionnels à τ.

$$\tau = \frac{\text{temps (T.S.)}\ N_f\text{(T.S.)}}{\text{temps (Cowell)}\ N_f\text{(Cowell)}} ,$$

l'ordre de chaque méthode ayant été optimisé dans chaque cas afin de minimiser le produit en question (pour la précision demandée), et N_f désignant le nombre d'évaluations de la fonction perturbatrice.

References

1 Kovalevsky, J.: Revue générale des méthodes automatiques de calcul des théories analytiques de satellites artificiels. XXIIe. Congrès Internat. d'Astronautique, Bruxelles, September 1971.

2 Danby, J.M.A.: Matrix perturbation methods using regularized coordinates. Celestial Mechanics 2 (1970) 311-318.

3 Broucke, R.: Perturbations in rectangular coordinates by iteration. Celestial Mechanics 1 (1969) 110-126.

4 Ceschino, F.; Kuntzmann, J.: Problèmes différentiels de conditions initiales. Paris: Dunod Ed. 1963.

5 Gorn, S.; Moore, R.: Report n. 893, Ballistic Research Labs., Aberdeen Proving Ground, Maryland, 1953.

6 Roy, A.E.; Moran, P.E.; Black, W.: Studies in the application of recurrence relations to special perturbation methods. I. Celestial Mechanics 6 (1972) 468-482.

7 Balmino, G.: Intégration numérique des équations de Lagrange. Applications diverses. Rapport interne G.R.G.S. Observatoire de Meudon, Février 1972.

8 Henrici, P.: Discrete Variable Methods in Ordinary Differential Equations. New York: J. Wiley and Sons, 1962.

9 Black, W.: Practical considerations in the numerical solution of the N-body problem by Taylor series methods. Celestial Mechanics 8 (1973) 357-370.

10 Brouwer, D.: Astronomical Journal 46 (1937) 149.

11 Kovalevsky, J.; Levallois, J.J.: Géodésie Générale, Tome IV. Mécanique Céleste. Paris: Eyrolles Ed. 1971.

12 Bettis, D.G.: A Runge-Kutta-Nyström algorithm. Celestial Mechanics 8 (1973) 229-293.

13 Fehlberg, E.: Classical Eighth and lower order Runge-Kutta Formulas with Stepsize Control for Special Second Order Differential Equations. NASA. TR-R-381, 1972.

14 Hull, T.E. et al.: Technical Report n. 29. Department of Computer Science, University of Toronto, 1972.

15 Moran, P.E. et al.: Studies in the application of recurrence relations to special perturbations methods, IV. Celestial Mechanics 8 (1973) 405-428.

16 Lapidus, L.; Seinfeld, J.: Numerical Solution of Ordinary Differential Equations, New York, 1971.

17 Butcher, J.C.: On Runge-Kutta processes of high order. J. Austral. Math. Soc. 4 (1964) 179-194.

18 Shanks, E.B.: Solutions of differential equations by evaluations of functions. J. Austral. Math. Soc. 20 (1966) 21-38.

19 Moran, P.E.: Studies in the application of recurrence relations to special perturbations methods. II. Celestial Mechanics 7 (1973) 122-135.

20 Sheffield, C.H.: Generalized Multi-step methods with an application to orbit computation. Celestial Mechanics 1 (1969) 46-58.

21 Kulikov, D.K.: Integration of equations of Celestial Mechanics by Cowell's method with variable intervals. Dynamics of Satellites, IUTAM Symposium, Paris 1962.

22 Balmino, G.: Etude et programmation de la méthode d'intégration numérique de Cowell-Kulikov. Rapport de D.E.A., Observatoire de Meudon, Paris 1967.

23 Chassaing, J.P.; Roumiguier, A.: Description d'un programme de correction différentielle utilisant l'intégration numérique. Rapport interne CNES. JPCH/AR/0.339/CB/MT, 1970.

24 Moynot, B.: Caractéristiques d'un programme de correction différentielle en géodésie dynamique. Astronomical Research 1971, 27-35, D. Reidel Pub. Co.

25 Oesterwinter, C.; Cohen, C.H.: New orbital elements for moon and planets. Celestial Mechanics 5 (1972) 317-395.

26 Bulirsch, R.; Stoer, J.: Asymptotic upper and lower bounds for results of extrapolation methods. Numerische Mathematik 8 (1966) 93-104.

27 Bulirsch, R.; Stoer, J.: Numerical treatment of ordinary differential equations by extrapolation methods. Numerische Mathematik 8 (1966) 93-104.

28 Gragg, W.: Repeated extrapolation to the limit in the numerical solution of ordinary differential equations. Thesis, UCLA, 1963.

29 Gautschi, W.: Numerical integration of ordinary differential equations based on trigonometric polynomials. Numerische Mathematik 3 (1961).

30 Brock, P.; Murray, F.J.: The use of exponential sums in step-by-step integration. M.T.A.C. 6 (1952) 63-78, 138-150.

31 Stiefel, E.: Remarks on numerical integration of Keplerian orbits. Celestial Mechanics 2 (1970) 274-281.

32 Bettis, D.G.: Stabilization of finite difference methods of numerical integration. Celestial Mechanics 2 (1970).

33 Hull, T.E.; Enright, W.H.; Fellen, B.M.; Sedgwick, A.E.: Comparing numerical methods for ordinary differential equations. SIAM, J. Numerical Analysis, v. 9. n. 4 (Dec. 1972).

34 Gantmacher, F.R.: Théorie des Matrices. Paris: Dunod, Ed. 1960.

35 Frazer, R.A. et al.: Elementary Matrices. Cambridge University Press 1963.

36 Ditto, F.H.: Partial derivatives used in trajectory estimation. Celestial Mechanics 1 (1969) 130-140.

37 Uphoff, C.: Numerical averaging in orbit prediction. AIAA Journal, v. 11, n. 11 (1973).

38 Balmino, G.: Evaluation à long terme de faibles perturbations orbitales par quadratures numériques. Rapport interne G.R.G.S., CNES, 1974, en préparation.

39 Roth, E.A.: Fast computation of high eccentricity orbits by the stroboscopic method. Celestial Mechanics 8 (1973) 245-249.

J-Adaptive Estimation with Estimated Noise Statistics *

A. H. JAZWINSKI** and C. HIPKINS***
Business and Technological Systems, Inc.
Seabrook, Maryland 20801 USA

Abstract

The previously developed J-Adaptive sequential estimator (1,2) is extended to include simultaneous estimation of the noise statistics in a statistical model for unmodeled system dynamics. This extension completely automates the estimator, eliminating the requirement of an analyst in the loop. Simulations in satellite orbit determination demonstrate the efficacy of the sequential estimation algorithm.

Introduction

In previous works (1,2) the authors have described a sequential estimator (called the J-Adaptive estimator) which tracks the state of a dynamical system with a simplified system model by simultaneously tracking a forcing term which approximates the unmodeled system dynamics. This estimator has been very successfully simulated in satellite orbit determination problems (1), where the unmodeled system dynamics result from unmodeled earth oblateness accelerations. The J-Adaptive estimator requires, however, the specification of a statistical model for the unmodeled system dynamics. The present study extends the J-Adaptive estimator to include the simultaneous estimation of the statistics of the unmodeled system accelerations, thus completely automating the estimation process.

---

*Research supported by the National Aeronautics and Space Administration, Goddard Space Flight Center, under Contract No. NAS 5-22144.

**President

***Vice President

The basic concept behind the J-Adaptive estimator (1, 2,5) is as follows. The real dynamical system

$$\ddot{R} = f(R,\dot{R}) \quad , \tag{1}$$

where R is the satellite position vector ($\dot{R} = V$), is modeled in the estimator as

$$\ddot{R} = f_m(R,\dot{R}) + S(t)u \quad , \tag{2}$$

so that $S(t)u$ approximates

$$f(R,\dot{R}) - f_m(R,\dot{R}) \tag{3}$$

which is the modeling error made in the estimator dynamics. $S(t)$ is a specified "symmetry" matrix and u is the unmodeled acceleration vector. The J-Adaptive estimator sequentially tracks the state R,V and the unmodeled acceleration vector u from tracking (or other) data. In the estimator, the unmodeled acceleration u is modeled statistically as a random polynomial in time.

Estimation of Noise Statistics

The discrete statistical model selected here for the unmodeled acceleration vector u is

$$u(k+1) = u(k) + \tau w(k) \tag{4}$$

where $\{w(k)\}$ is a 3-vector, zero-mean, white Gaussian sequence with identically distributed components and covariance matrix

$$\varepsilon\{w(k)w^T(\ell)\} = q(k)I\ \delta_{k\ell} \quad , \tag{5}$$

and where $q(k)$ will be estimated from the data via the adaptive filtering approach of (3,4). The complete dynamical model (in discrete form with the index k denoting the discrete measurement time instants) is therefore given by

$$x(k+1) = \phi_m(x(k), Su(t)) \quad ,$$
$$u(t) = u(k) + (t-t_k)w(k) \quad ,$$
$$u(k+1) = u(k) + \tau w(k) \tag{6}$$

where x is the state vector (R,V) and ϕ_m is the mapping defined by Eq. (2). The measurement model is given by

$$y_i(k) = h_i(x(k)) + v_i(k) \quad , \quad i = 1...N \tag{7}$$

where y_i are the scalar measurements available at time k, and $\{v_i(k)\}$ are independent, scalar, zero-mean, Gaussian noise sequences with

$$\varepsilon\{v_i(k)v_i^T(\ell)\} = R_{ii}(k)\delta_{k\ell} \quad . \tag{8}$$

The measurement model is written in vector form as

$$y(k) = h(x(k)) + v(k) \quad , \tag{9}$$

where

$$y^T = (y_1 \cdots y_N) \quad ,$$
$$h^T = (h_1 \cdots h_N) \quad ,$$
$$R = \text{diag}\ (R_{ii}) \quad . \tag{10}$$

Now define the sensitivity matrices

$$\Phi(k+1,k) = \left[\frac{\partial x(k+1)}{\partial x(k)}\right] = \left[\frac{\partial \phi_m}{\partial x(k)}\right]$$

$$\Psi(k+1,k) = \left[\frac{\partial x(k+1)}{\partial u(k)}\right] \triangleq \begin{bmatrix} \frac{\tau^2}{2} & S \\ \tau & S \end{bmatrix}$$

$$\Gamma(k+1,k) = \left[\frac{\partial x(k+1)}{\partial w(k)}\right] \triangleq \begin{bmatrix} \frac{\tau^3}{6} & S \\ \frac{\tau^2}{2} & S \end{bmatrix}$$

$$M(k) = \begin{bmatrix} M_1(k) \\ \vdots \\ M_N(k) \end{bmatrix} = \left[\frac{\partial h(x(k))}{\partial x(k)}\right] \tag{11}$$

and the covariance and correlation matrices

$$P(k|\ell) = \varepsilon\{[x(k)-\hat{x}(k|\ell)]\ [x(k)-\hat{x}(k|\ell)]^T\}$$
$$C_u(k|\ell) = \varepsilon\{[x(k)-\hat{x}(k|\ell)]\ [u(k)-\hat{u}(k|\ell)]^T\}$$
$$U(k|\ell) = \varepsilon\{[u(k)-\hat{u}(k|\ell)]\ [u(k)-\hat{u}(k|\ell)]^T\} \tag{12}$$

where $P(k|\ell)$ is the state estimation error covariance matrix (covariance of errors in $\hat{x}(k|\ell)$) at time t_k, given all measurement up to and including time t_ℓ. C_u and U have similar meaning. Then the usual extended Kalman filter (4) with variance q(k) known and specified is given below.

Prediction (in time) is given by

$$\hat{x}(k+1|k) = \phi_m(\hat{x}(k|k), S\hat{u}(t|k))$$
$$\hat{u}(k+1|k) = \hat{u}(k|k) \tag{13}$$

$$P'(k+1\ k) = \Phi P(k|k)\Phi^T + \Phi C_u(k|k)\Psi^T + \Psi C_u^T(k|k)\Phi^T + \Psi U(k|k)\Psi^T$$
$$P(k+1|k) = P'(k+1|k)+q(k)\Gamma\Gamma^T$$
$$C_u(k+1|k) = \Phi C_u(k|k)+\Psi U(k|k)+q(k)\tau\Gamma$$
$$U(k+1|k) = U(k|k)+q(k)\tau^2\Gamma \tag{14}$$

where the argument (k+1,k) has been omitted from the matrices Φ, Ψ and Γ. The (measurement) update is given by

$$\hat{x}(k|k) = \hat{x}(k|k-1)+K_x(k)\ [y(k)-h(\hat{x}(k|k-1))]$$
$$\hat{u}(k|k) = \hat{u}(k|k-1)+K_u(k)\ [y(k)-h(\hat{x}(k|k-1))] \tag{15}$$

$$P(k|k) = P(k|k-1)-K_x(k)M(k)P(k|k-1)$$
$$C_u(k|k) = C_u(k|k-1)-K_x(k)M(k)C_u(k|k-1)$$
$$U(k|k) = U(k|k-1)-K_u(k)M(k)C_u(k|k-1) \tag{16}$$

where

$$K_x(k) = P(k|k-1)M^T(k)Y^{-1}(k)$$
$$K_u(k) = C_u^T(k|k-1)M^T(k)Y^{-1}(k) \tag{17}$$

and where

$$Y(k) = M(k)P(k|k-1)M^T(k)+R(k)\quad . \tag{18}$$

Now the variance q(k) will be estimated via the techniques of Refs. (3,4). To that end, define the average normalized predicted residual at time k+1,

$$\bar{r}_{k+1} = \frac{1}{N} \sum_{\ell=1}^{N} \frac{[y_\ell(k+1) - h_\ell(x(k+1|k))]}{R_{\ell\ell}^{1/2}(k+1)} \quad . \tag{19}$$

It is easy to compute

$$\varepsilon\{\bar{r}_{k+1}\} = \varepsilon\{\bar{r}_{k+1}^2 | q(k) \equiv 0\} + dq(k) \tag{20}$$

where

$$\varepsilon\{r_{k+1}^2 | q(k) \equiv 0\} =$$

$$\frac{1}{N} + \frac{1}{N^2} \sum_{\ell=1}^{N} \sum_{m=1}^{N} \frac{M_\ell(k+1)P'(k+1|k)M_m^T(k+1)}{R_{\ell\ell}^{1/2}(k+1)\ R_{mm}^{1/2}(k+1)} \tag{21}$$

$$d = \frac{1}{N^2} \sum_{\ell=1}^{N} \sum_{m=1}^{N} \frac{M_\ell(k+1)\Gamma\Gamma^T M_m^T(k+1)}{R_{\ell\ell}^{1/2}(k+1)\ R_m^{1/2}(k+1)} \quad . \tag{22}$$

Then the most probable q(k) based on $\bar{r}_{k+1}$ is given by (3,4)

$$\bar{q}(k) = \begin{cases} \frac{1}{d}\left[\bar{r}_{k+1}^2 - \alpha\varepsilon\{\bar{r}_{k+1}^2 | q(k) \equiv 0\}\right] & \text{, if positive} \\ 0 & \text{, otherwise} \end{cases} \tag{23}$$

with $\alpha = 1$.

The estimate $\hat{q}(k)$ of q(k) utilized is an exponentially age-weighted average of $\bar{q}(k)$ generated from

$$c_k = ac_{k-1} + 1 \quad , \quad c_{-1} = 0 \quad , \quad a < 1 \quad ,$$

$$\hat{q}(k) = \frac{c_k^{-1}}{c_k} \hat{q}(k-1) + \frac{1}{c_k} \bar{q}(k) \quad ,$$

$$\hat{q}(-1) = 0 \quad . \tag{24}$$

The $\hat{q}(k)$ from Eqs. (24) is used for $q(k)$ in the filter equations (14).

To complete the estimator equations, the symmetry matrix S in Eqs. (11) and (13) must be specified. A matrix S used with success is given by

$$S(k) = \frac{1}{|\hat{R}(k|k)|} \begin{bmatrix} -\hat{x}(k|k) & 0 & 0 \\ 0 & -\hat{y}(k|k) & 0 \\ 0 & 0 & -\hat{z}(k|k) \end{bmatrix} . \quad (25)$$

This points the acceleration vector u at the center of the earth.

Simulations

Simulations presented below feature a circular, polar orbit of approximately 1000 km altitude. The "real" system of force model consists of two-body forces plus a 4×4 harmonic field. The estimator dynamical model is two-body. Tracking (range and range rate) is simulated from several tracking stations to give continuous, and often overlapping, station coverage, sampled every 5 sec. Tracking begins at the north pole and continues for 50 minutes, which is slightly less than one-half the orbital period. The error models for range and range rate measurements are simplified, consisting only of additive white Gaussian noise with standard deviations σ_ρ = 3m, $\sigma_{\dot{\rho}}$ = 1cm/sec. Figures 1-7 show the estimator performance.

Figures 1-3 show the tracking of the unmodeled acceleration. The solid lines represent the actual unmodeled acceleration, while the dots are the estimates $S\hat{u}$. The z-component of the unmodeled acceleration dominates, being of the order of 10^{-5} (initially 10^{-4}) km/sec^2, and is therefore tracked most precisely. The x and y acceleration component estimates are more noisy, but are clearly tracked. Note that there is an initial transient lasting about 300 sec or so before the estimator locks onto the unmodeled acceleration. Figures 4 and 5 show the normalized measurement residuals (normalized by their noise standard deviations) for range and range rate. The initial transient described above exhibits itself in the range rate residuals, producing, initially, rather large residuals. This transient dies out, however, and the measurement residuals then appear quite white and are within the measurement noise level. The absence of residuals after about 2000 sec of tracking is merely a programming problem in the computer-generated plots; the residuals continue to exhibit appropriate statistical properties for the balance of 1000 sec. Figures 6 and 7 show the actual state estimation errors in position and velocity; namely $|\hat{R}-R|$ and $|\hat{V}-V|$. It is seen that after an initial transient the state estimation errors are quite comparable with the measurement noise. That is, position is estimated to about 3m and velocity to about 1-5 cm/sec.

Consider the initial (~ 300 sec) transient which is observed in Figures 1-7. This transient is, in part, due to the transient in $\hat{u}$; that is, the state cannot be estimated precisely until unmodeled acceleration is. This is not the whole story, however. Note that the range rate normalized residuals, while sometimes as large as 5σ initially, are random and cannot explain the large initial state estimation errors, particularly in velocity. It is noted that the drop in state errors at about 300 sec coincides with acquisition of the satellite by a second tracking station. It is also noted that the position and velocity estimation errors are largely in the x and y components; the z component errors are quite small. This is simply due to the station-satellite geometry. Thus from the first station, x, y, $\dot{x}$ and $\dot{y}$ are not as observable as z and $\dot{z}$ are. This geometric observability clearly has an impact on the estimation of Su as well; and also on the estimation of $\hat{q}$. $\hat{q}$ guarantees residuals consistent with a projection of the state covariance matrix on the "available" measurement space; not on the unobservable subspace of state space. This last observation was well evidenced in a simulation with range rate measurements only. In that case, range rate residuals were just as nice as in the present simulation, while state errors were not quite as good during the first 300 sec of tracking. $\hat{q}$ compensated only for the state subspace it could see.

The interested reader may find additional simulations and discussions in Ref. (5).

Conclusions

Simulations demonstrate that the J-Adaptive estimator with estimated noise statistics can automatically, without human intervention, and with a very simple orbit model (e.g. two-body model), estimate satellite orbits to an accuracy comparable with the data noise levels, when excellent, continuous tracking coverage is available. Such tracking coverage will be available from satellite-to-satellite tracking. The fact that a two-body model can be utilized in the estimator prediction equation eliminates the need for numerical integration of the orbit dynamics equations, which in turn results in a very fast (computational) orbit estimation algorithm.

This estimator also tracks (estimates) the unmodeled acceleration vector. From these latter estimates a parametric model of the unmodeled accelerations may be recovered (2). The precision of tracking of these unmodeled accelerations is, of course, a function of the measurement noise levels (measurement precision).

References

1 Jazwinski, A.H.; Hipkins, C.: Adaptive Estimation for Model Enhancement in Satellite Orbit Determination. Business and Technological Systems, Inc., BTS-TR-72-1, July 1972.

2 Jazwinski, A.H.: Adaptive Sequential Estimation with Applications. Presented at the Fifth IFAC Symposium on Automatic Control in Space, Genoa, Italy, June 1973. Also Automatica 10 (1974) 203-207.

3 Jazwinski, A.H.: Adaptive Filtering. Automatica 5 (1969) 475-485.

4 Jazwinski, A.H.: Stochastic Processes and Filtering Theory. New York: Academic Press, Inc. 1970.

5 Jazwinski, A.H.; Hipkins, C.: J-Adaptive Estimation with Estimated Noise Statistics. Business and Technological Systems, Inc., BTS-TR-73-5, November 1973.

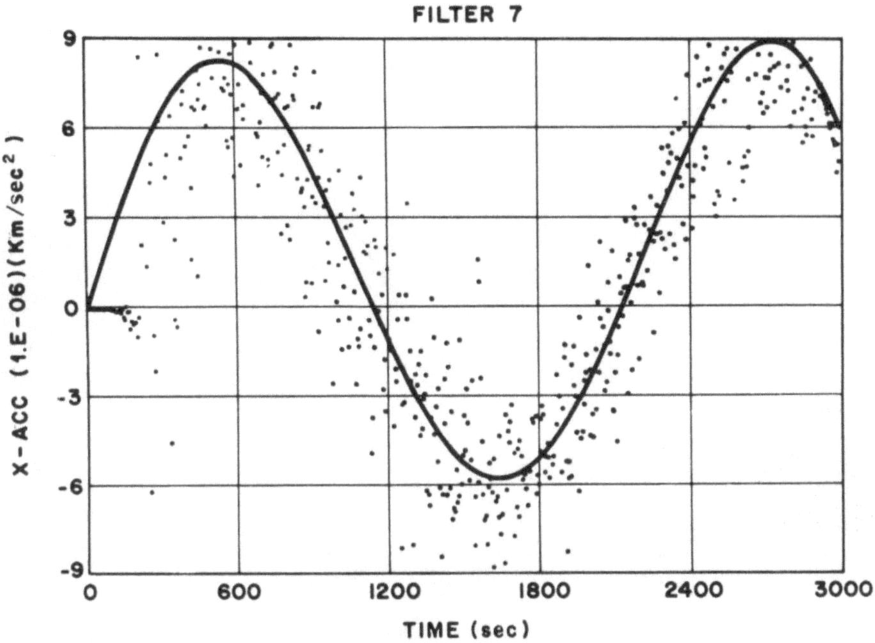

Figure 1. Tracking the x-acceleration.

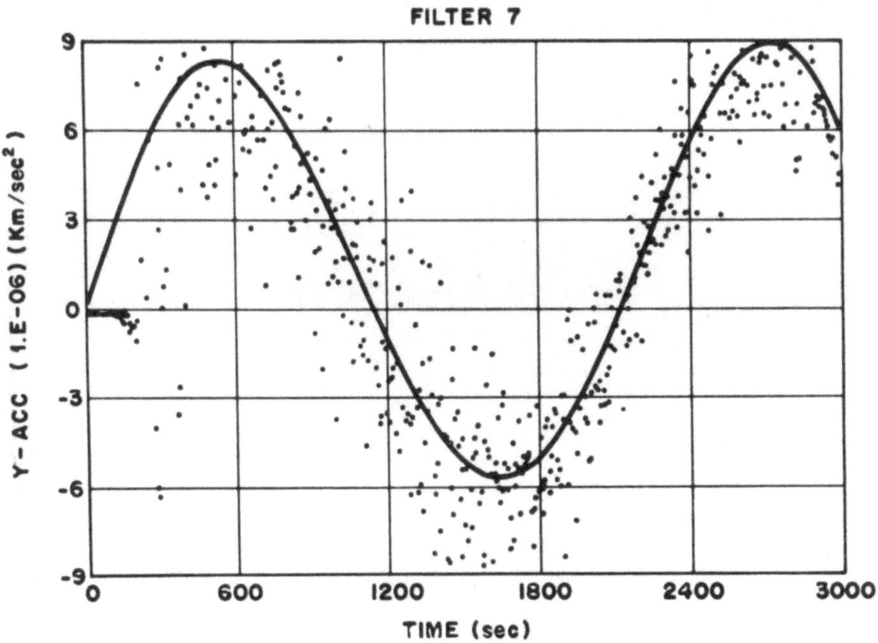

Figure 2. Tracking the y-acceleration.

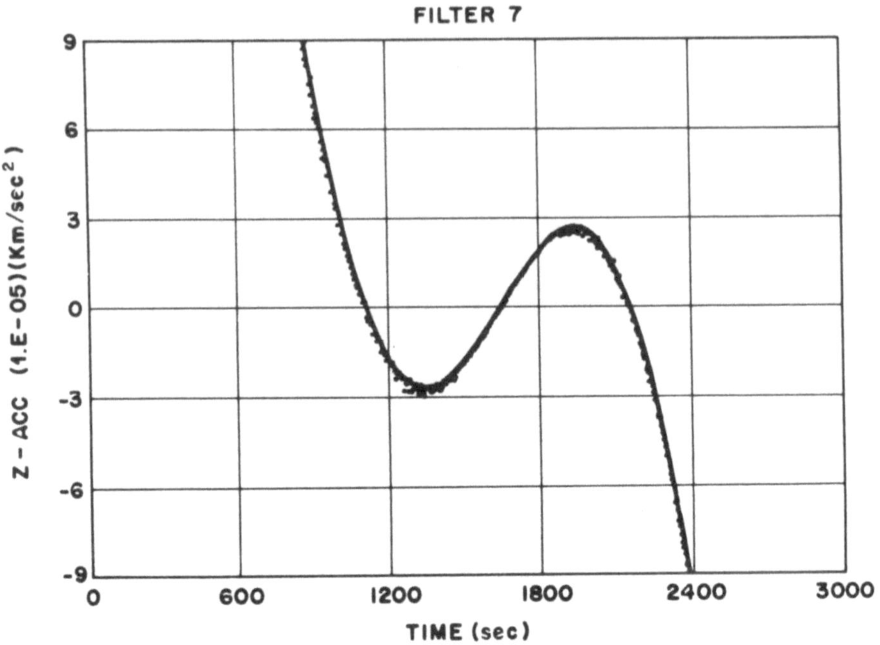

Figure 3. Tracking the z-acceleration.

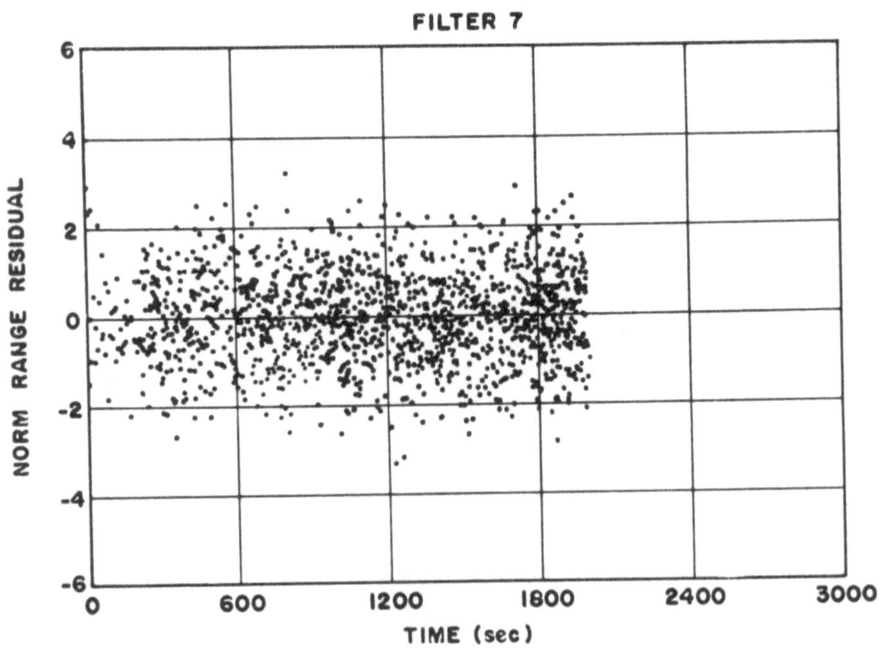

Figure 4. Normalized range residuals.

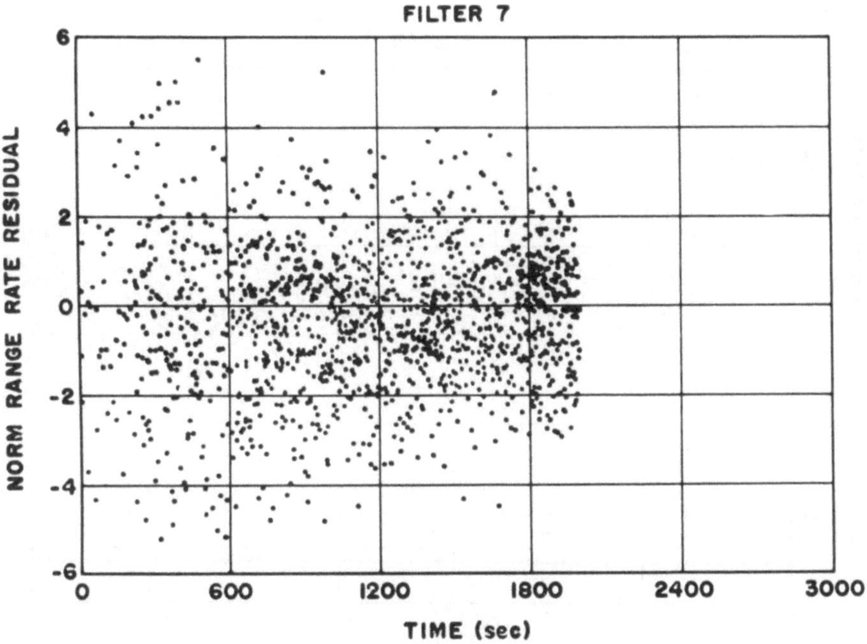

Figure 5. Normalized range rate residuals.

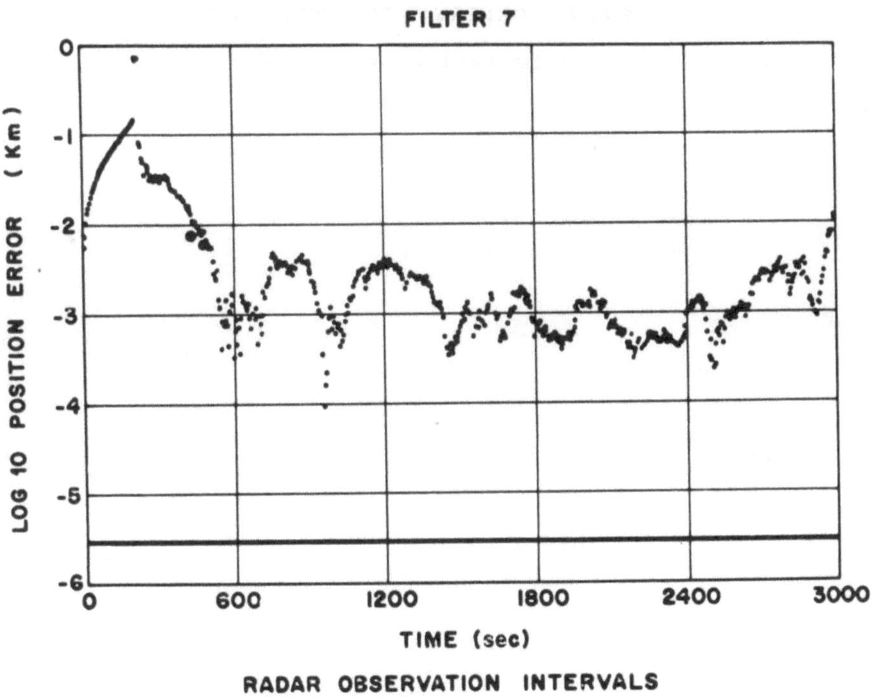

Figure 6. Position estimation errors.

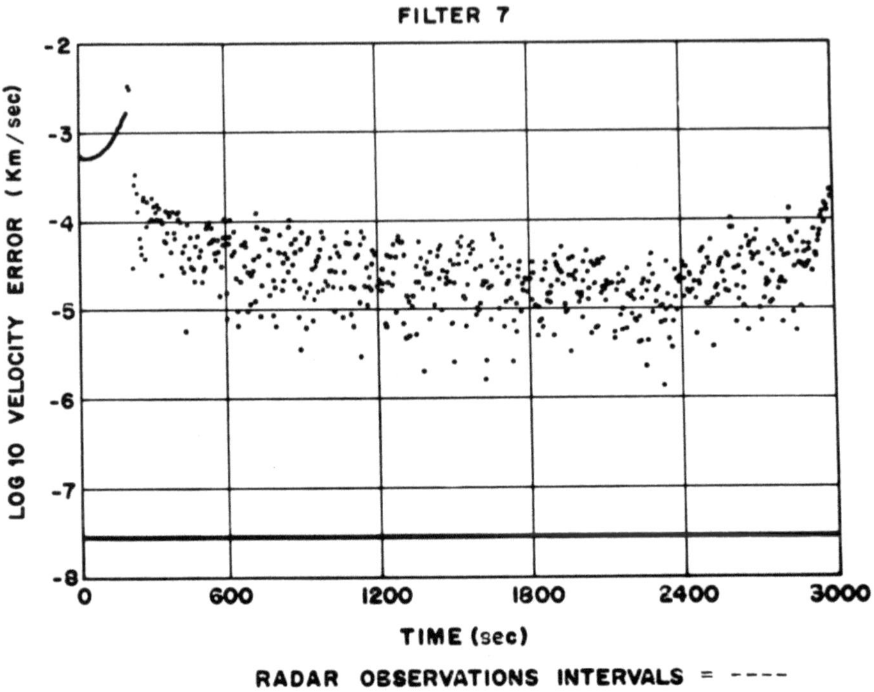

Figure 7. Velocity estimation errors.

Importance of the Coupling Effects Between Earth Potential Harmonics in the Motion of an Artificial Satellite. Computed and Checked Solution of This Coupling Problem. The Case of J_7

X. BERGER
G.R.G.S.
Meudon Observatory, France

Abstract

The "coupling" effect arising from the simultaneous presence of different harmonics can be great compared with the direct effect of the harmonics. The coupling effect J_2-J_7 is very significant. Theoretical, numerical and graphic studies have demonstrated this importance and given the reason for it. We have expressed literally the coupling expressions for the zonal and tesseral, nonresonant, harmonics (development of coupling equations, integration, verification by analytical-numerical comparisons, studies of the form and properties of the terms...). We indicate which coupling effects must be considered for a good representation of the motion of any satellite, and for the determination of zonal harmonics. All this has been done and has been established by computer checking.

Last year an analytical theory of the motion of an artificial satellite in the Earth's gravity field, developed literally to the third order, was presented (1). The literal expressions of the effects of the harmonics on the elements of any satellite have been established, as well as the literal expressions of the coupling effects between the harmonics. 200 million coupling terms have been determined and have finally given, after simplification, only some hundreds. We have found terms in J_n (i.e. coupling terms in J_nJ_2/J_2; J_2 disappears), J_n^2/J_2^2, J_nJ_2, J_n^2/J_2, J_nJ_m/J_2^2... The solution has been checked by comparisons between analytical theory and numerical integration. The accuracy is 30 cm for any orbit with eccentricity not too small.

Here we wish to point out the importance of these expressions, that is the terms additional to the direct integration of Lagrange's equations, not only for a good description of the motion of a satellite, but also for the determination of zonal harmonics.

The literal development of any term is in alternating series (opposite signs for the numerical coefficients) for the powers of the sine-cosine of inclination. The developments to the second and third orders of the theory, and the four iterations of the problem, necessary to obtain a complete solution, decrease the effect of the eccentricity. This, and the large number of terms coupling between themselves, increases the numerical coefficients in the literal expressions of the terms. Consequently, we find coupling terms as great, and sometimes greater than the terms of the direct solution of Lagrange's equations.

It is impossible to say before the literal development is obtained, which coupling effect is important for a given satellite. But the lower the eccentricity, the larger is the effect of the second and third order terms, and also of the coupling terms (the reason for this lies in the fact that the effect of the eccentricity decreases by a factor 2 with each order in the literal development). Of course, we are not referring here, in this text, to the well-known terms in J_n/J_2, coming from the coupling effect between the secular terms in J_2 and the long period terms in and i in J_n/J_2. Evidently, these terms are too important to be forgotten in any solution. To give an idea, consider these examples, and also the Tables and Figures 1 to 4:

D1C satellite (e = 0.084): the coupling effects J_2-J_5, J_2-J_6, J_2-J_8, J_2-J_9 are small and almost negligible. But the J_2-J_7 coupling effect is large, four times J_2-J_4, 40 m in amplitude (Figs. 1 and 2).

GEOS B satellite (e = 0.031): the long period terms in J_2^2 are as large as the terms in J_2: 800 m in amplitude (Table 1); the long period terms in J_3^2/J_2^2 in 2ω are four times greater than the long period terms in J_2: 3 km in amplitude; the long period terms in J_3J_5/J_2^2 in 2ω are equal to one half the long period terms in J_2.

PEOLE satellite (e = 0.016): the long period terms in J_2^2 are five times greater than the long period terms in J_2...(Table 2).

In Figures 3-5 one can see the growing effect of the long and short period terms in J_2^3 (the analytical theory includes all the periodic terms in J_2, J_2^2 and secular terms in J_2, J_2^2, J_2^3) with decrease in the eccentricity, especially the effect of the long period term in $\underline{a}$ in J_2^3.

The same type of determination of amplitudes and periods can be made for any coupling effect, and for any satellite (Table 3). This shows that the expressions given by the direct integration of Lagrange's equations, with the long period coupling terms in J_n/J_2 included, are not sufficient to give a correct description of the main effects. Consequently, it is necessary to complete the equations for the determination of zonal harmonics. It is possible that the values of the zonal coefficients will be altered by 40 percent with such new correct equations; this work is in progress.

We have observed very important coupling terms with J_7. This results from the very large numerical coefficients in the literal developments. For the other harmonics, the alternating series do not give very large numerical coefficients in the solution. For J_7, the 4 million non-negligible coupling effects which have been literally resolved have produced very important coefficients; the alternating form of the starting developments used does not lead here to normal coefficients. Thus, we are sure to meet very large coupling effects with this harmonic for any satellite. In Fig. 2, one can see the verification of the exact determination of the literal expressions for the coupling effect J_2-J_7.

The coupling expressions between any 2 or 3 harmonics need to be considered for a correct description of the motion of a satellite. For the determination of zonal harmonics, the most useful coupling effects are:

$$J_2^2,\ J_2\text{-}J_3,\ J_2\text{-}J_4,\ J_3\text{-}J_5,\ J_2\text{-}J_7,\ J_3\text{-}J_7 \quad .$$

We have also developed the coupling effects for tesseral harmonics, but we have found only a small effect at mean period on the angular elements. We have expressed this literally. It is non-negligible only for harmonics of low degree (C_{22}, S_{22},...) and has to be considered only for a very good description of the motion of a satellite, and not for the determination of tesseral harmonics (Fig. 6).

For resonant tesseral harmonics, we have developed the coupling equations (a large number: 27) and we are now solving the two most important ones: one concerning a long period effect in $J_{nq}J_{mq}/J_2$ (n=m or n≠m) (and one more on M in $J_{nq}J_{mq}/J_2^2$), and one concerning a mean period effect in $J_{nq}J_2$. The long period effect is as large as the effect of the long period resonant terms, coming from the direct integration of Lagrange's equations. Thus, the expressions of these coupling effects must be considered, not only for the determination of the resonant tesseral harmonics, but also for the determination of zonal harmonics, because the periods are exactly 1ω, 2ω, 3ω,... (Fig. 7).

In the determination of harmonics, the importance of coupling effects lies also in the fact that the equations are not linear with respect to the harmonics. This is a very important consideration.

In conclusion, it must be stressed that the actual determinations of zonal or resonant harmonics by using analytical theories are not quite satisfactory (4-7) in regard to the present precision of the observations. It is necessary to modify the equations by taking into account the coupling terms; the literal expressions of these coupling terms have been determined (2-3). The analytical theory has been checked by comparisons with numerical integration for each coupling effect, and the accuracy of the expressions for the coupling terms and the need for their adoption is considered to be established.

Table 1. Influence on the elements e, Ω, ω of some long period terms in the development of the analytical theory. Amplitudes (in km, radian) and periods of these terms for the GEOS B satellite.

Nature of the term	Period	Amplitude		
		e	Ω	ω
term in J_2	2ω	0.380×10^{-5}	0.324×10^{-6}	0.121×10^{-3}
term in J_2^2	2ω	0.179×10^{-5}	0.464×10^{-9}	0.114×10^{-3}
coupling term in J_3J_2/J_2^2 and coupling term in J_3J_2/J_2	1ω	0.541×10^{-6}	0.126×10^{-3}	0.978×10^{-4}
coupling term in J_3^2/J_2^2	2ω	0.694×10^{-5}	0.	0.449×10^{-3}
coupling term in J_3J_2/J_2	3ω	0.579×10^{-7}	0.	0.369×10^{-5}
coupling term in J_3J_5/J_2^2	2ω	0.688×10^{-7}	0.	0.758×10^{-4}
coupling term in J_7J_2/J_2	1ω	0.109×10^{-5}	0.633×10^{-7}	0.261×10^{-3}
coupling term in J_7^2/J_2^2	2ω	0.532×10^{-6}	0.	0.392×10^{-3}

Table 2. Influence on the elements e, Ω, ω of some long period terms in the development of the analytical theory. Amplitudes (in km, radian) and periods of these terms for the PEOLE satellite.

Nature of the term	Period	Amplitude		
		e	Ω	ω
term in J_2	2ω	0.405×10^{-6}	0.972×10^{-7}	0.245×10^{-4}
term in J_2^2	2ω	0.833×10^{-6}	0.107×10^{-8}	0.100×10^{-3}
coupling term in J_3J_2/J_2^2 and coupling term in J_3J_2/J_2	1ω	0.359×10^{-6}	0.121×10^{-4}	0.790×10^{-5}
coupling term in J_3^2/J_2^2	2ω	0.115×10^{-5}	0.	0.140×10^{-3}
coupling term in J_3J_5/J_2^2	2ω	0.367×10^{-6}	0.	0.446×10^{-4}
coupling term in J_7J_2/J_2	1ω	0.724×10^{-6}	0.	0.441×10^{-4}
coupling term in J_7^2/J_2^2	2ω	0.913×10^{-7}	0.	0.111×10^{-4}

Table 3. Influence on the elements e, Ω, ω of some long period terms in the development of the analytical theory. Amplitudes (in radian) and periods of these terms for the Echo 1 rocket.

Nature of the term	Period	Amplitude		
		e	Ω	ω
term in J_2	2ω	0.193×10^{-5}	0.380×10^{-7}	0.175×10^{-3}
term in J_2^2	2ω	0.128×10^{-5}	0.608×10^{-10}	0.232×10^{-3}
coupling term in J_3J_2/J_2^2 and coupling term in J_3J_2/J_2	1ω	0.642×10^{-6}	0.393×10^{-4}	0.693×10^{-5}
coupling term in J_3^2/J_2^2	2ω	0.107×10^{-4}	0.	0.195×10^{-2}
coupling term in J_3J_2/J_2	3ω	0.604×10^{-7}	0.	0.110×10^{-4}
coupling term in J_3J_4/J_2^2	1ω	0.465×10^{-6}	0.172×10^{-6}	0.339×10^{-4}
	3ω	0.928×10^{-7}	0.	0.170×10^{-4}
coupling term in J_3J_5/J_2^2	2ω	0.118×10^{-5}	0.	0.216×10^{-3}

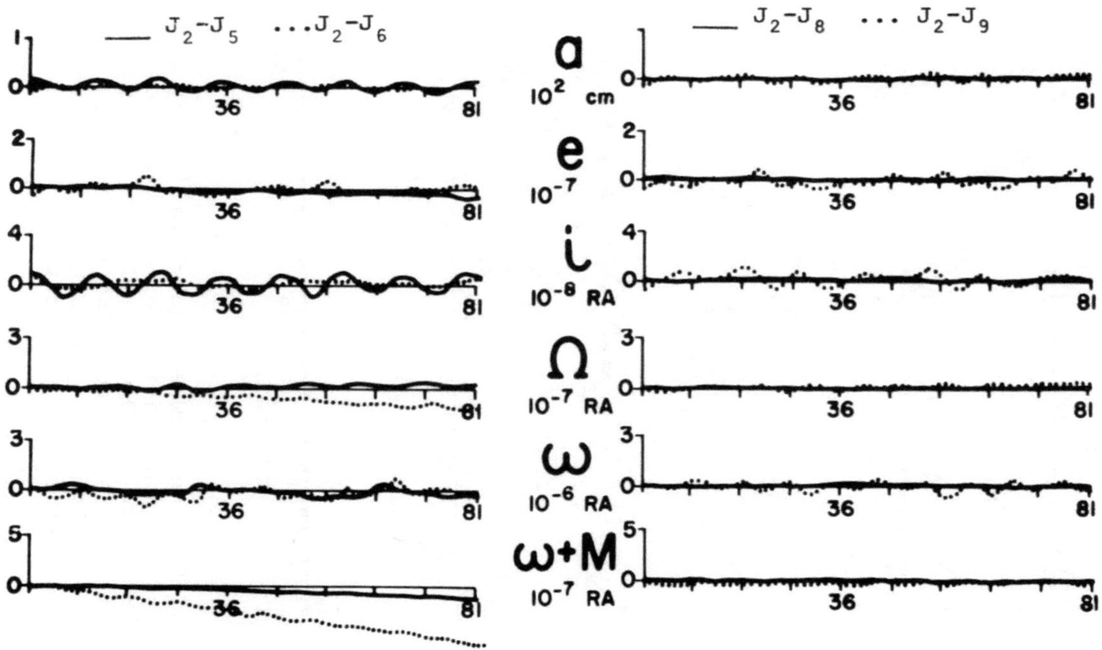

Figure 1. Difference between analytical theory and numerical integration for the influences of J_2-J_n on the D1C satellite. The analytical theory does not contain the coupling terms, except the terms in J_n/J_2. Abscissa: time in hours.

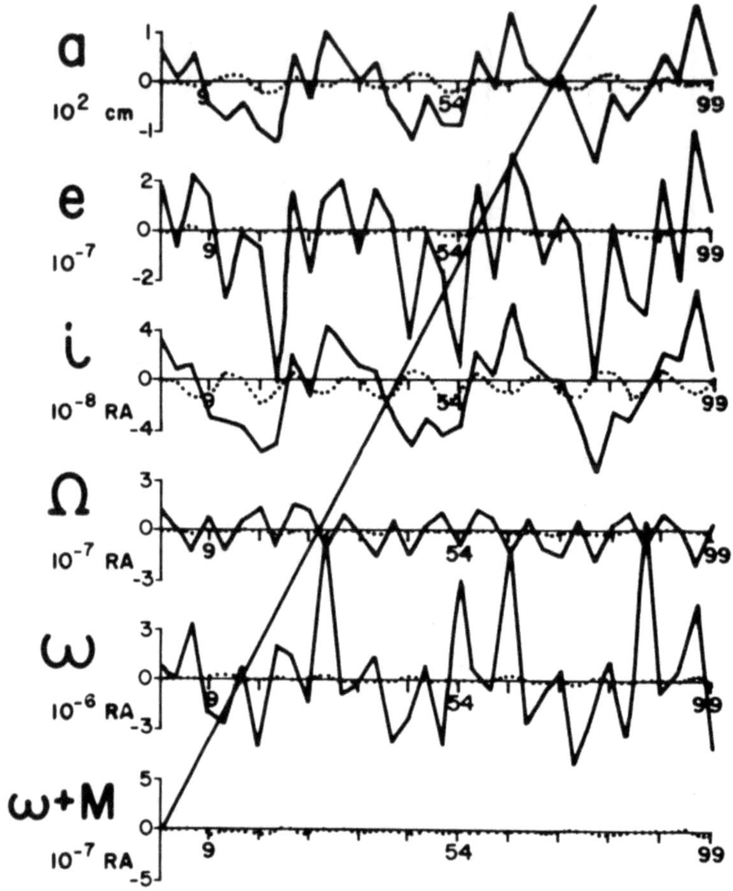

Figure 2. Difference between analytical theory and numerical integration for the influence of J_2 and J_7 on the D1C satellite. Abscissa: time in hours. ___ coupling effect J_2-J_7 (the analytical theory does not contain the coupling terms, except the terms in J_7/J_2); ... with all the coupling terms in the analytical theory.

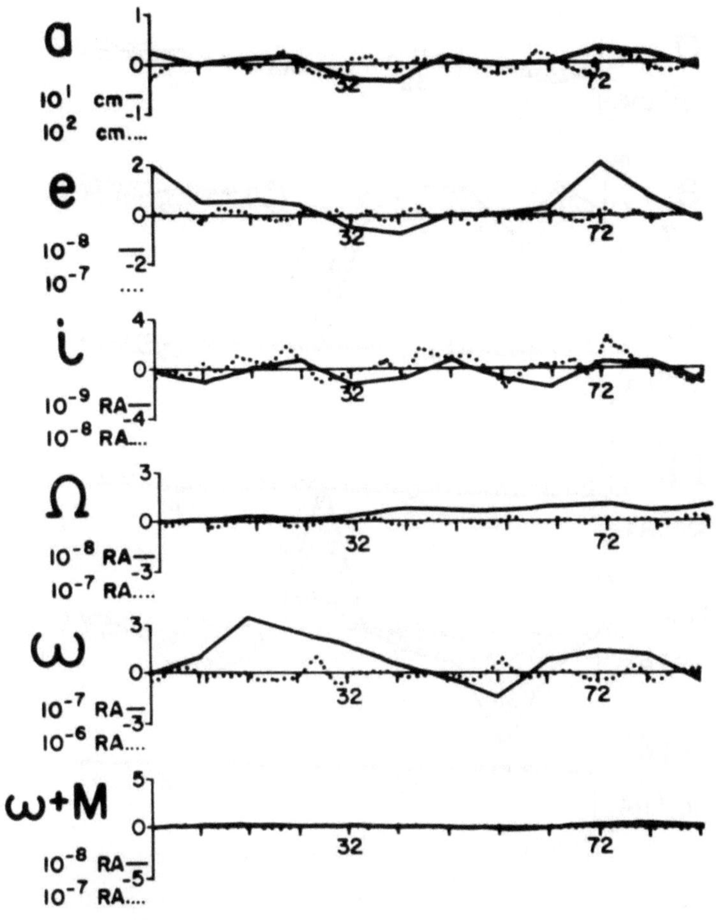

Figure 3. Difference between analytical theory and numerical integration for the D1C satellite (e = 0.08). Effect of the missing terms and of the wrong terms in the analytical theory. Abscissa: time in hours. ___ theory in J_2; ... theory in J_2-J_3-J_4-J_5 with the coupling effects J_2-J_3, J_2-J_4, J_2-J_5, J_3-J_4, J_3-J_5, J_4-J_5 in the analytical theory.

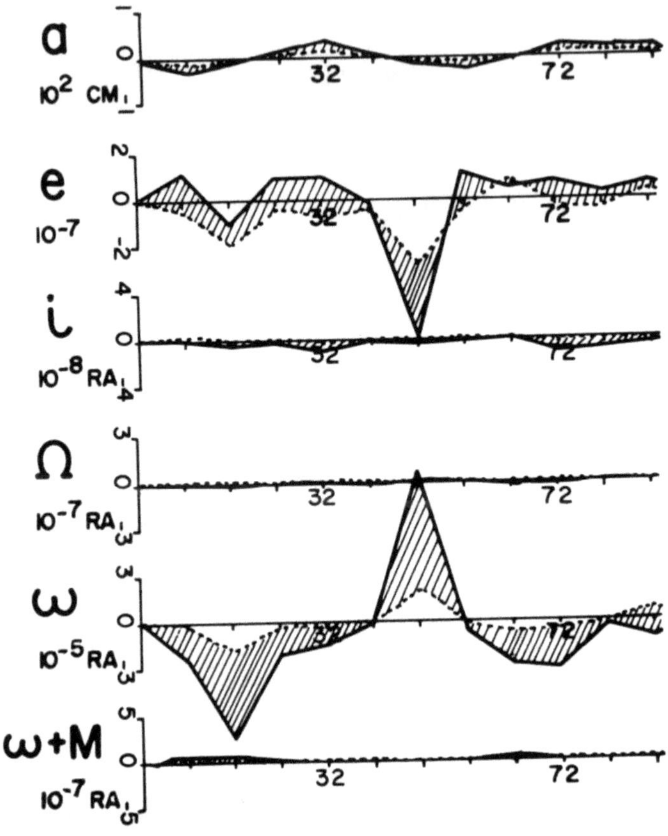

Figure 4. Difference between analytical theory and numerical integration for GEOS B satellite ($e = 0.031$). Effect of the missing terms (mainly the long period terms in J_2^3 and the coupling terms of the fourth order) in the analytical theory. Abscissa: time in hours. ... theory in J_2; ___ theory in J_2-J_3.

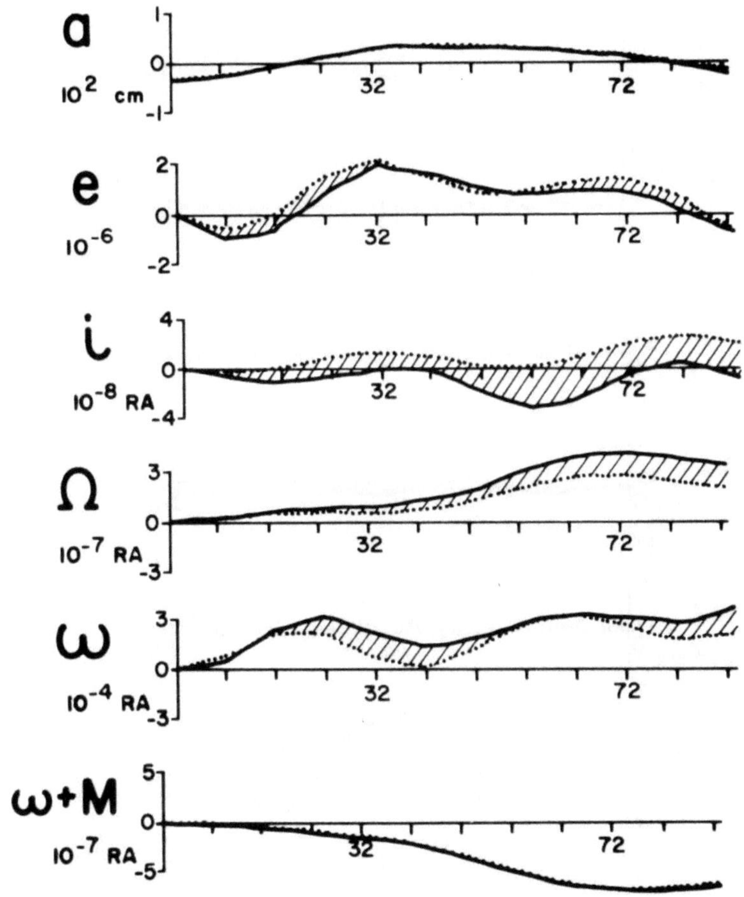

Figure 5. Difference between analytical theory and numerical integration for PEOLE satellite ($e = 0.016$). Effect of the missing terms (mainly the long period terms in J_2^3) in the analytical theory. Abscissa: time in hours. ... theory in J_2; ___ theory in J_2-J_3.

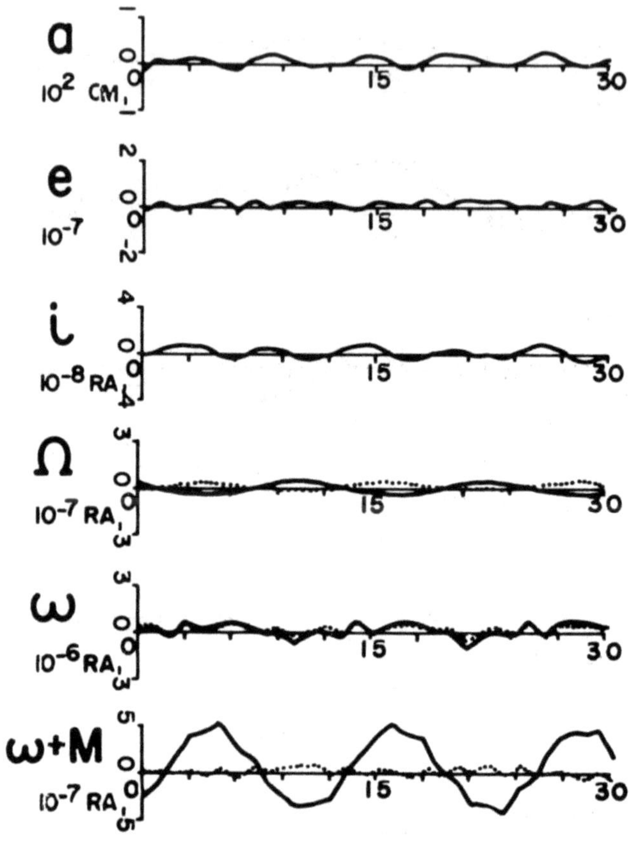

Figure 6. Difference between analytical theory and numerical integration for the influence of J_2, C_{22}, S_{22} on D1C satellite. Abscissa: time in hours. ___ with no coupling terms J_2-C_{22}, J_2-S_{22}, C_{22}-S_{22} in the analytical theory; ... with these coupling terms in the analytical theory.

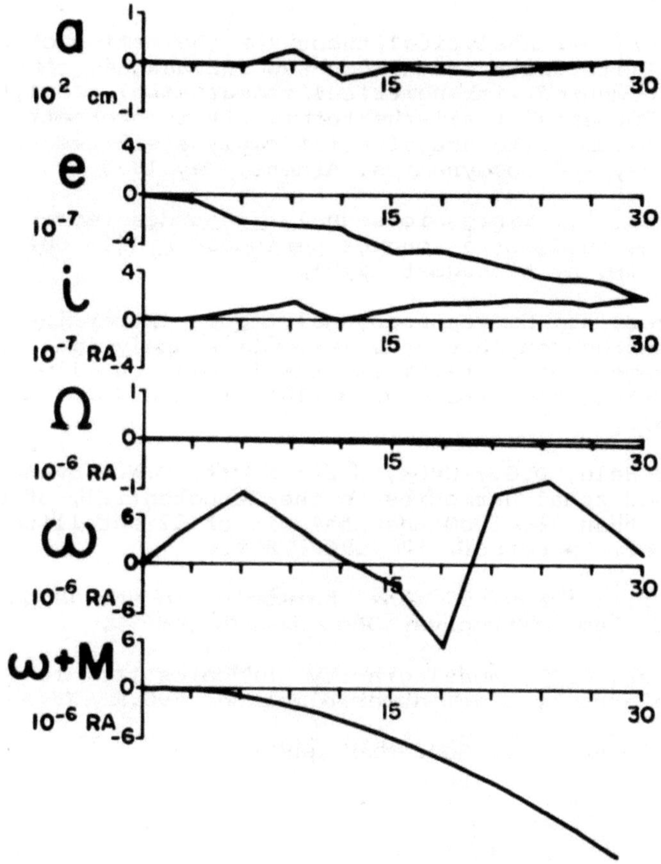

Figure 7. Difference between analytical theory and numerical integration for the influence of J_2 and $C_{13,13}$ on GEOS B satellite. The analytical theory does not contain the coupling terms between the two harmonics. Abscissa: time in hours.

References

1 Berger, X.: Analytical theory of the motion of a satellite under earth-moon-sun influences, computed and compared with numerical integration. Analytical developments of all the terms. First International Symposium - The use of artificial satellites for Geodesy and Geodynamics. Athens. May 1973.

2 Berger, X.: Expressions analytiques des termes seculaires en J_2^3 et à longues periodes en J_2^2. Bulletin G.R.G.S. n. 5 (August 1972).

3 Berger, X.: Expressions analytiques littérales des termes de couplage dans la théorie analytique du mouvement des satellites artificiels sous l'action du potentiel terrestre. Bulletin G.R.G.S. n. 8 (June 1973).

4 King Hele, D.G.; Cook, G.E.; Scott, D.W.: Evaluation of odd zonal harmonics in the Geopotential, of degree less than 33, from the analysis of 22 satellite orbits. Planet. Space Sci. 17 (1969) 629.

5 Kozai, Y.: Revised zonal harmonics in the geopotential. Spec. Rep. Astrophys. Obs. n. 295 (1969).

6 Wagner, C.A.: Zonal gravity harmonics from long satellite arcs by a seminumeric method. J.G.R. n. 17 (1973).

7 Kovalevsky, J.: Géodésie Genérale 4 (1971).

On the Comparison of Numerical Theories of Orbital Motion

J. DERRAL MULHOLLAND
Groupe de Recherche de Géodésie Spatiale, Meudon, France
and The University of Texas at Austin U.S.A.

Abstract

The numerical study of the motion of the closest natural satellite, the Moon, provides a means to examine the general concept of a numerical theory, as opposed to a simple ephemeris. Attempts to reconcile a discrepancy between our calculations and those of Oesterwinter and Cohen permit some conclusions on the documentation required for such a theory to be complete.

Introduction

The topics covered by this symposium include "computer applications" and "numerical theories" in the orbital dynamics of satellites. One may justly ask what is the difference. What is to be understood by the term "numerical theory" if it is to be a useful concept? I propose to examine this question in the context of a study of our own natural satellite, the Moon, and to offer some conclusions.

The numerical integration of the equations of motion for systems of celestial bodies has become both practical and fashionable in recent years. Within its capacity, the computer doesn't care how many bodies are in the system, nor (within some large domain) what their initial conditions are. There is no need to linearize or to arrange by orders of a small parameter. Explicit truncation decisions are avoided and critical arguments are largely irrelevant. The computer simply accepts the input and returns an ephemeris. This process is so cheap that, for reasonable-size systems and time spans, one can produce numerous ephemerides based on various initial conditions, to examine the effect of various changes directly in the coordinates. By contrast, the production of a new ephemeris through revision or generation of an analytic theory is still a monumental task. When faced with the present inadequacy of available analytic theories, one is not surprised that numerical integration has been used widely, not only for the ephemeris requirements of spacecraft navigation, but also for more direct studies of the physics of the planets and satellites (1), the foundations of dynamical theory (2), and various aspects of stellar and galactic astronomy. Indeed, the AENA ephemerides (3,4) of the giant planets have been so founded since 1960. It has sometimes been suggested that numerical integrations be made the basis for all working ephemerides. Some few have even claimed that a conventional theory could be established by the adoption of an acceleration model and a set of arbitrary constants. At this point, it is necessary to make a clear distinction between an ephemeris and a theory of motion, a distinction that will be valid for both analytic and numerical constructions.

An ephemeris is the practical realization of a theory of motion, consisting of a tabular set of coordinate values covering some finite interval sufficiently densely for reliable interpolation. It may be provisional or private and totally undocumented. We are not concerned here with ephemerides, but with the underlying theories, the means by which an ephemeris can be replicated by someone other than its author. A theory of motion for an object consists of the collection of means sufficient to define and permit the calculation of a unique set of coordinates at any arbitrarily-selected epoch. In the particular case that we call an analytic theory, this is provided by a set of algebraic relations giving some set of coordinates as a function of time, together with the arbitrary constants of the theory. There is no general agreement, however, on the proper constitution of a numerical theory. To find one, it is necessary to ask what is required for the independent numerical replication of an ephemeris generated by numerical integration.

A Comparison of Numerical Theories

A theory of motion can be compared with observations or with other theories. Discussion of the observations has the purpose of establishing the "best values" of the arbitrary constants for use with that theory, a process that is identical in principle for both analytic and numerical theories; it is not of interest to this discussion. The comparison of theories may be desired to establish the relations between sets of arbitrary constants, to evaluate the differences in physical models or in construction techniques. For this reason, it is customary to publish such details of an analytic theory as will permit the examination of these fundamentals, as well as the easy replication of an ephemeris. The situation is not so straightforward for numerical theories, and consequently the norm for documentation has not been very high. In this case, ephemeris replication is not assured, and indeed this may be the principal object of the comparison. An illustration of the problem has been provided by my attempt to compare two numerical integrations of the lunar orbit.

The numerical theory of the Moon and planets by Oesterwinter and Cohen (5) is the most carefully and completely documented one known to me. They have described minutely the formulation of their acceleration model, the scheme of numerical integration, the arbitrary constants, derivation of the partial derivatives, the method for solving the normal equations, and the statistics relating to the final solution. The osculating state vectors were given at the epoch and at three additional dates. An admirably thorough work, but disturbing in having noted an unresolved disparity between the results of their STRETCH program and some computations of mine. Consequently, I undertook to duplicate the ephemeris of the Oesterwinter-Cohen theory to identify the discrepancy, if it still existed, and to examine the effects of certain details of their acceleration model.

The computer program used in this study is the 11-body solar system integrator originally designed by C. L. Lawson at the Jet Propulsion Laboratory. The integration algorithm is an Adams-Moulton predictor-corrector very similar to that of Oesterwinter and Cohen, with nearly all operations performed in double precision to control the loss of end figure significance in long integrations. It has been used with the IBM 7094 (16 digits), UNIVAC 1108 (18 digits), and presently a CDC 6600/6400 (29 digits) with no inter-machine discrepancies at an unexpected level.

It was not difficult to produce a numerical integration based on the same constants and initial conditions, but this was not in fact very useful. The two programs had used different metrics for the post-

Newtonian equations of motion, so we produced strikingly different solar systems. It was necessary to transform their state vectors into our coordinate frame to produce comparable results. This was avoided by a modification to our program to permit use of the same geometry. With this change, acceptable agreement was obtained for all objects except Earth and Moon. For these, the seemingly minor differences of formulation are significant, because of the Moon's large geocentric mean motion and the higher accuracy required for its geocentric coordinates. The effects of some of these differences on the epoch acceleration are given below.

Table 1. Effects of model differences between Oesterwinter-Cohen integration and that of present study, lunar geocentric acceleration at the epoch JED 2441200.5

Nature of difference	defect source	digits of agreement
Earth equator unprecessed, J_2 effect	OC	10
Solar effect on Earth J_2 neglected	M	11
Incomplete rotation of lunar equator	OC	11
Solar effect on Moon J_2 neglected	M	13
Precession neglected, lunar J_2 effect	OC	13

Various other differences produced changes that were insignificant relative to the 14 digit precision of the STRETCH results. After these differences were accounted for, it was apparent that even the Earth-Moon barycenter motion was satisfactory, but that the geocentric motion of the Moon was not. This was discouraging, since our program treats the Moon as a special case, while the other does not. It appeared likely that the longitude error of 0".25 after 800 days was ours and that it existed even in the Newtonian point-mass integration. With the aid of Dr. Oesterwinter, we tried to find the cause. The mathematical formulations and the integration starting procedures were examined with great care and found to be compatible, but there was a difference in the 12th figure of the lunar epoch accelerations. Before the reason for this could be determined, the STRETCH computer died an administrative death and was dismantled, preventing a definitive resolution of the problem. Some circumstantial evidence exists, however. A careful manual calculation by Oesterwinter (6) of the Earth-Moon-Sun contributions and a CDC 6600 computation by Oesterwinter (7) of the planetary perturbations, when combined, give a total geocentric lunar acceleration that differs from ours only in the 14th digit. Comparisons with a STRETCH three-body computation indicate that the acceleration degradation is about equally divided between the main problem and the planetary part. Thus, it appears that the problem lay in electromechanical differences between the computers. I now believe, but cannot prove, that the STRETCH program produced excessive integration errors in the lunar geocentric orbit.

Lessons from Experience

Perhaps the importance of this study is not so much the apparent result as the implication that follows from the difficulties encountered in reaching it. On a purely operational level, certain control information was not available for the STRETCH computations, because the programmer failed to print a l l of the input data on the output listing, so that some conditions could not be verified at a later date. One extra printed page costs less than a milliday of a scientist's time, after all, which may be about the minimum time required to determine that the information does not exist. At a more substantive level, it was not possible to trace the exact problem because of calculations that were unavailable -- the starting table, total accelerations, acceleration components.

The widespread and successful use of numerical integration for ephemerides has led Deprit and others to insist on the comparison of numerical theories. This might be done simply by a direct exchange and differencing of the ephemerides, but this does not seem an adequate test of a "theory". One must attempt replication in order to understand and accept fully its validity "as a theory". Documentation of the theory must include sufficient information to measure whether replication has been accomplished, and if not, why not. I propose here a number of criteria toward this end; they have not all been met by any numerical theory known to me.

First, I propose to define one form of a numerical theory of motion to be a working program, "together with its computer," the initial conditions at the epoch, the values of all physical parameters that influence the computation of the motion, and the program control information used in generating the ephemeris. This will have a utility to the limited circle who have access to an identical machine and for whom the machine is not a consciously relevant factor. The Oesterwinter-Cohen paper is an excellent model for documentation of this form, but with the computer gone, the theory is no longer complete.

The second, and more permanent form of a numerical theory should consist of the program, the numerical and control information, plus enough supplementary information to permit one to know "what was really done by the computer," both the effects of modelling and of finite word length. I suggest that this include 1) a step-by-step ephemeris of position, velocity, and acceleration for some interval including but exceeding the integration starting table; 2) a detailed decomposition of the epoch accelerations into their individual contributions from different effects; and 3) an ephemeris adequate for detailed application or comparison. Until this level of specification is achieved, the numerical theory will never be quite so respectable as analytic theory.

Acknowledgements

I am deeply indebted to Claus Oesterwinter for his cheerful cooperation with this study, including supplementary computer runs. This research was supported in part by the National Aeronautics and Space Administration (USA) under grant NGR 44-012-219 at the University of Texas, and by the Centre National de la Recherche Scientifique (France).

References

1 Bender, P.L.; Currie, D.G.; Dicke, R.H.; Eckhardt, D.H.; Faller, J.E.; Kaula, W.M.; Mulholland, J.D.; Plotkin, H.H.; Poultney, S.K.; Silverberg, E.C.; Wilkinson, D.T.; Williams, J.G.; Alley, C.O.: The Lunar Laser Ranging Experiment. Science 182 (1973) 229-238.

2 Van Flandern, T.C.: A Determination of the Rate of Change of G. Presented to the meeting of the American Geophysical Union (1974).

3 American Ephemeris and Nautical Almanac. Published annually by U.S. Government Printing Office, Washington, D.C.

4 Eckert, W.J.; Brouwer, D.; Clemence, G.M.; Coordinates of the Five Outer Planets 1653-2060. Astr. Papers Am. Ephem. 12 (1951) U.S. Government Printing Office, Washington, D.C.

5 Oesterwinter, C.; Cohen, C.J.: New Orbital Elements for Moon and Planets. Cel. Mech. 5 (1972) 317-395.

6 Oesterwinter, C.: private communication, April 18, 1973.

7 Oesterwinter, C.: private communication, July 3, 1973.

Stabilization and Real World Satellite Problem

C. E. VELEZ
NASA/Goddard Space Flight Center
Greenbelt, Maryland, USA

Abstract

Many recent investigations have concerned themselves with the development of transformations which tend to "dynamically stabilize", "regularize", or otherwise "smooth" the classical Newtonian equations of unperturbed motion with the goal of improving the numerical behavior of the system for the perturbed case. Examples of such transformations include stabilization by the application of integrals, independent variable transformations, and dependent variable transformations for the case of highly perturbed motion. The specific transformations examined include the energy control term and element equations developed by Stiefel, Scheifele and Baumgarte, and time regularization techniques.

It is shown by computer simulations of several NASA satellite orbits that in many cases, while such transformations do tend to improve the error growth properties of the perturbed problem, they do so at an added expense which degrades the overall efficiency. It is further shown that this results from adverse effects these transformations have on (i) the numerical stability regions of the integrator, or (ii) local error uniformity, or (iii) formulation "overhead". It is concluded that while such transformations may allow one to attain high precision results which are otherwise unattainable in classical formulations, with the exception of analytic stepsize control, they offer little or no advantage over the classical methods in the intermediate accuracy range, when overall efficiency is considered.

Introduction

As the physical models of the perturbing accelerations acting on an artificial satellite become more complex, and as greater needs for precision orbit determination and prediction from applications such as satellite geodesy arise, standard numerical approaches become very costly and of questionable validity in high accuracy regions. As a result there has been a growing interest in the literature in transformations of the orbital equations or motion which are said to "stabilize" or "uniformize" the dynamical system with the hopes of either reducing the accumulation of local numerical errors, or allowing the use of larger stepsizes in the transformed space, or both. Examples of such transformations include the use of new independent variables - time regularization (1), transformations to orbital parameter space which tend to decouple fast and slow variables (2), or reduce the equations to nearly linear form (3), and the use of integrals as control terms (5). The purpose of this study is to evaluate the relative merits of such transformations and to attempt to isolate the difficulties associated with them.

The basic criterion to be used to conclude relative merit should be discussed. Generally the question to be asked is: "Can I achieve equal precision in the transformed space with less computer cost?" However one cannot talk of computer efficiency independently of the choice of numerical integration scheme used, since in many cases transformations affect the performance of the numerical technique itself (1), i.e., the interplay between round-off errors, various truncation errors and stability regions of numerical integrators are affected by choice of formulation, and therefore can cause confusion as to what formulation gives best results. In this study we have used high-order multistep integration schemes based on the Newtonian interpolating polynomials frequently labeled the Störmer-Cowell or Adams-Bashforth, Adams-Moulton methods. These were selected because our studies have shown them to be most efficient for the integration of the classical Newtonian equations (6). Hence, to the above question should be added the proviso: "with the given high-order multistep integrator."

Along the same lines, a frequently used efficiency gauge is the number of function evaluations needed to achieve a given accuracy. In situations where computer cost per step is dominated by the algebraic expressions in the perturbing accelerations this may be realistic. However, many of the transformations proposed are quite complex, frequently involving a higher dimensional state space, so that the cost of transforming back and forth between the Cartesian space and element space may be an important factor in overall efficiency. In such cases, actual computer cost is the only possible reliable gauge, although this also involves factors such as programming practices, reliability of computer clocks, etc. Because of such elements of vagueness, care should be taken in drawing conclusions from such numerical studies and results should serve primarily as qualitative indications rather than quantitative. With these provisos in mind, the following sections will describe the various transformations tested and the experimental results.

A cursory examination of these results will show that there is no universal formulation that is best for all circumstances. It seems that an improvement occurs only when there is a physically or mathematically understandable reason why one would expect improved efficiency from a new formulation. More will be said of this below.

Formulations Tested

I. Classical Cowell Formulation (C)

The conventional Cartesian formulation is given by

$$\ddot{x} = \frac{\mu x}{r^3} + P(t,x,\dot{x}) \qquad (1)$$

where P is the sum of all perturbative accelerations

acting on the satellite. It is noted that in this formulation the equation is Lyapunov unstable (L-unstable) even for the case P = 0, so that errors committed during the integration of Eq. (1) will grow quadratically with time (1). Further, local truncation errors are inversely proportional to r, so that the uniformization of local truncation errors over an orbital period for high eccentricity orbits - a highly desirable feature - must be done with a numerical stepsize control. Such controls generally require estimates of local errors and may involve costly re-start procedures. A remedy for some of these difficulties is given by the time regularized formulation.

II. Time Regularized Formulation (TR)

This formulation involves a transformation of the independent variable t to a new variable s defined by the relation

$$\frac{dt}{ds} = \frac{r^n}{\sqrt{\mu}} \quad . \tag{2}$$

The spatial equations transform to

$$x'' = \frac{nr'x'}{r} - r^{2n-3}x + \frac{r^{2n}}{\mu} P(t,x,\dot{x}) \tag{3}$$

where ' denotes differentiation with respect to the new variable s. The state space now includes t as a dependent variable. This mechanism has proved to be quite effective in uniformizing the local error, allowing the use of fixed stepsize integrators for cases of highly elliptic motion. However this uniformization is controlled by the parameter n which is highly dependent on the perturbation model P and the state space elements. In Cartesian space the value of n = 1 uniformizes the case where P = 0, whereas n = 1.5 performs similarly for the case where P contains the J_2 perturbations (1). Experience has shown that the optimum value of n for more general perturbations is near 1.5. However in element formulations such as the Lagrange equations the situation is different. Here uniformization occurs at larger values of n such as n = 2. The reason for this is fairly clear if one notes that as n varies from 1 to 2, the distribution of function evaluations around the orbit becomes more dense in the perigee region and less dense elsewhere. In Cartesian space the two-body acceleration is still significant at apogee and hence function evaluations must be made in this region to model this force accurately. On the other hand, element formulations solve the two-body problem exactly, so that in situations where the apogee region is nearly a conic, very large stepsizes can be taken. Therefore, in such cases, uniformization occurs when the function evaluations are concentrated at perigee. As will be seen below most element formulations tested included a time regularization of n = 1 or 2.

It is noted that the change of independent variable reduces the L-instability of the spatial Eqs. (3) at the expense of equivalent instability resulting from Eq. (2). Improvements in the integration of Eq. (2) may be made by the use of time elements (3) which allow the exact integration of this equation for Keplerian motion and reduce the instability for the case of perturbed motion (7). Time elements are included in the KS and DS formulations described below.

III. Stabilization by the Use of Integrals

A series of methods have evolved centered around the concept that integrals or near-integrals of the motion may be used to reduce the accumulation of truncation errors in Eq. (3) thereby allowing the use of larger stepsizes to achieve the same accuracy. The Baumgarte approach (B) described in Ref. 4 involves the addition of a control term (CT) to Eq. (3) of the form

$$CT_B = \alpha x \left[\frac{x'^2}{2r} - \frac{1}{r} + \frac{V+h}{\mu}\right] \tag{4}$$

where V is the perturbing potential energy of the system evaluated at a given state, and h is the negative of the total energy satisfying

$$h' = - x' \cdot P - \frac{x}{\sqrt{\mu}} \left[\frac{\partial V}{\partial t}\right] \quad . \tag{5}$$

For the case $V = P = 0$ and $\alpha = 1$, the resulting equation is L-stable but is unstable otherwise. It is also noted that this control term does not force the solution onto the correct energy surface (8), i.e., $CT_B \not\to 0$ as $t \to \infty$. This can be seen by multiplication of the Baumgarte equation in time domain by $\dot{x}$ leading to the equation

$$\frac{d}{dt} \ln \left(CT_B\right) = - \frac{d}{dt}\left(\ln r\right) \tag{6}$$

which has the solution

$$r\left(CT_B\right) = \text{constant} \quad ,$$

so that numerical errors in the control term are maintained. In contrast, a control term presented by Baumgarte and Steifel (BS) in Ref. 9 has the property that it exponentially decays with time. It is of the form

$$CT_{BS} = \frac{\gamma \dot{x}}{\dot{x}^2} \left[\frac{\dot{x}^2}{2} - \frac{1}{r} + V + h\right] \quad , \tag{7}$$

As with the Baumgarte control term, for the case $V = P = 0$ and any γ, the resulting equation is L-stable but is unstable otherwise.

An approach based on a similar principle has been proposed by Nacozy (10). In his case the equations of motion are not altered, but rather the error in the energy is used to correct the state at each step of the integration process, in effect returning the state to the correct energy surface like the BS formulation.

The principle of virtual work is used to correct the position of the satellite, i.e., if δh is the observed error in energy, then the correction δx in state is given by

$$\delta x = \frac{\delta h \ddot{x}}{|\ddot{x}|^2} \tag{8}$$

The idea here is that although the equations of motion remain unstable in the L-sense, the instability is not permitted to cause a rapid accumulation of local errors.

As far as L-stabilization is concerned, it is noted that to achieve L-stability one must remove the period as a degree of freedom of the dynamical system. The use of the energy integral as an a-priori constant achieves this for two-body motion, and one can see that all the processes which stabilize use such a-priori information. This feature also exists in most variations of parameter techniques discussed below.

IV. Stabilization by the Use of Elements

This represents a large class of formulations based on the idea of using an analytic reference orbit and numerically integrating variations in the parameters due to additional perturbations. Most of these formulations are L-stable in the unperturbed case since they use exact integrals as coordinates. Further, as will be seen below, such formulations frequently include a time regularization which partially uniformizes local errors. However, there are two disadvantages from a numerical point of view:

(1) Element formulations invariably require the integration of a system of first-order equations (of dimension 6 or higher) in contrast to the integration of a system of second order Eqs. (3) in rectangular coordinates. It is known (6) that integrators which approximate solutions to second-order systems (e.g. Cowell), are more efficient than their first order counterparts (e.g. Adams).

(2) Element formulations involve additional computations to transform from element space to Cartesian space during the computations. Sophisticated elements which transform the equations of motion to oscillators (e.g. KS) or which represent high order intermediate orbits (11) involve significant "overhead" which must be considered in evaluating efficiency. It is frequently true that although such formulations allow the use of larger stepsizes to achieve comparable accuracy, the cost per step is also higher, so that no overall efficiency gain is obtained.

In this study 3 element formulations were tested:

(1) A Gaussian formulation (VOP) based on equinoctial elements α defined by

$$\alpha = \begin{cases} a \\ e \sin (\omega+\Omega) \\ e \cos (\omega+\Omega) \\ \tan i/2 \sin \Omega \\ \tan i/2 \cos \Omega \\ E + \omega + \Omega \end{cases}$$

and where the differential equation is of the form

$$\dot{\alpha} = \frac{\partial \alpha}{\partial \dot{x}} \cdot P(t,x,\dot{x}) \quad . \tag{9}$$

(2) The Kustaanheimo-Stiefel (KS) element formulation
(3) consisting of:

- α, β - 4-dimensional vectors representing perturbations in the oscillator coefficients
- ω - perturbation in oscillator frequency
- τ - perturbation in eccentric anomaly time element.

In this formulation the generalized eccentric anomaly is the independent variable (corresponding to n = 1 in the TR formulation).

(3) Canonical Delaunay - Similar Elements (DS)
This formulation required the integration of variations in elements α, β with true anomaly as the independent variable. The variables are defined by (12,13)

$$\alpha = \begin{cases} \text{total conservative energy } L \\ \text{total angular momentum } G \\ \text{Z component of angular momentum } H \\ \text{perturbing energy} \end{cases}$$

$$\beta = \begin{cases} \text{"mean" mean anomaly } l \\ \text{argument of pericenter } g \\ \text{longitude of ascending node } h \\ \text{generalized true anomaly} \end{cases}$$

which are obtained by numerical integration of Gaussian equations together with a time element corrèsponding to the true anomaly. This formulation has numerical advantages in that it uses several near-integrals of the motion and hence is totally stable in the unperturbed case and nearly linearly stable in the perturbed case, but more importantly, the choice of independent variable is ideal for this type of formulation. The value n = 2 results in a highly dense distribution of function evaluations near perigee. This is desired since the principal perturbations occur in this region in element space, and therefore this choice uniformizes the local error for the oblateness problem. As will be seen below, these results are borne out by our numerical experiences.

Numerical Experiences

Before beginning this section I would like to reiterate the earlier caution that numerical experiences regarding relative performance of various formulations of the same problem are highly dependent on the "environment" in which the measurements are taken--numerical integrators, tolerances, programming techniques, etc. all influence the result, so that at best, one can only make conclusions within the context of the experiments, and at worst, the results are meager qualitative indications of relative merit.

Experiments were performed on three test orbits:

(1) GEOS-B : an intermediate distant near-circular orbit. Perturbations used include a full 15 x 15 geopotential, lunar and solar perturbations. Initial elements:

$a = 8080$	$\Omega = 337.2°$
$e = .07$	$\omega = 185.1°$
$i = 59.3°$	$M = 0$

(2) AE-C : a near-earth high-drag slightly eccentric orbit. Perturbations used include a 4 x 4 geopotential, drag, lunar and solar effects. Initial elements:

$a = 8525$	$\Omega = 91.5°$
$e = .23$	$\omega = 93.7°$
$i = 68°$	$M = 275°$

(3) SMS transfer : a highly eccentric orbit. Perturbations used were same as in the AE-C case. Initial elements:

$a = 24,548$	$\Omega = 260°$
$e = .73$	$\omega = 180°$
$i = 23.8°$	$M = 0°$

The results are displayed in two forms: a function evaluation versus accuracy graph, and tables containing computer time versus error in semi-major axis for selected formulations. On the graphs, accuracy is defined as $-\log_{10}$ (relative error) and is therefore a measure of significant digits. On such a graph the truncation error-limited portion is to the left. At smaller stepsizes, i.e., larger N, the accuracy approaches a limiting value and then decreases again in the roundoff error-limited region to the right.

Figures 1, 2 and 3 are accuracy graphs which display the results of integrating the test cases for approximately 50 revolutions using the Cowell (C), time regularized Cowell (TR), Kustaanheimo-Stiefel (KS), Baumgarte control term (B), Baumgarte-Stiefel control term (BS), Nacozy energy correction (N), variation of parameter using equinoctial elements (VOP), and Delaunay-Similar elements (DS) formulations.

An examination of these figures shows that the DS formulation gives the most efficient performance at all stepsizes for the three test orbits under the criterion that function evaluations are a measure of efficiency. This is especially true in the highly eccentric SMS case. Stabilization and the independent variable of this formulation explain this improvement. For the AE-C orbit (Fig. 2), where drag is a large perturbation, the "stabilization by use of integral" formulations are not effective against the timing (along track) errors caused by the nonconservative forces. In such cases the energy integral is nearly a "fast" variable and therefore suffers from rapid error growth as do the other coordinates in the perigee region. Again, the choice of true anomaly in the DS formulation results in a sampling of the drag region with higher frequency than the other formulations tested, resulting in improved performance. The effectiveness of good analytic stepsize control can be seen most dramatically in Fig. 3. Beyond the DS case ($n = 2$) the next most accurate formulations are those using $n = 1.5$. The KS theory, ($n = 1$), is next in the line of degrading formulations, followed by the $n = 0$ cases. It is clear that in element formulations the greatest errors occur at perigee, and one would do best with an independent variable commensurate with this situation.

Finally consider the GEOS results in Fig. 1. Because the orbit is nearly circular, the stepsize control is irrelevant. This graph shows the advantages of dynamic stabilization inherent in the VOP and integral formulations. Note that they are not as dramatic as in Fig. 3, so that one may infer that good stepsize control is more valuable than stabilization.

Tables 1, 2 and 3 show a comparison of the DS theory with Cowell or TR Cowell under the criterion of computer time (CPU). Here the picture as to what is most efficient changes slightly. One can see in Table 1 that for the GEOS orbit, the DS theory allows a 33% reduction in stepsize to achieve comparable accuracy, but the "overhead" of the formulation is approximately 50% so that no improvement in efficiency is obtained by using the DS formulation. In Table 2, for the AE-C case, the stepsize reduction allowed by DS is close to 50% so that in terms of CPU time, the two formulations perform nearly equal. Finally in Table 3, we see that the reduction is over 100%, so that significant improvement by use of the DS formulation is realized.

It is noted that these results do show that irrespective of efficiency considerations, stabilizing transformation allow the attainment of precisions which are unattainable with the Cowell formulations.

Conclusions

In conclusion, we have made the following observations with regard to the many experiments performed with various formulations and numerical integrations:

(1) Formulations of the satellite problem and integrators "interact" - one must be careful in making conclusions about the relative merits of one formulation over another independently of the numerical integrator used;

(2) Stabilizing transformations may give equal precision at larger stepsizes, but the "cost" inherent in the transformation may be an important factor in measuring efficiency;

(3) The proper choice of independent variable in element formulations is critical to good performance;

(4) In evaluating new formulations, the elements to be considered should include effects of:

- (a) "Uniformizing" the local error
- (b) "Smoothing" the derivatives
- (c) "Stabilizing" the equations
- (d) "Destabilizing" the integrator.

From our results the most we have concluded is that element formulations are valuable when the reference orbit is near the actual motion, and can yield very high precision results. Further, good analytic stepsize control is frequently the deciding factor as to the efficiency of a formulation, and of the formulations tested, an element formulation with $n = 2$ appears to afford some advantages.

Table 1

Satellite: GEOS-B Flight: 8 days
Forces: 4 x 4 Earth's Gravitational Field, Sun, and Moon
Semi-Major Axis at Final Time: 8067.0880332 km

Orbit Generator	Steps/rev.	CPU Time	Error in SMA
Cowell	100	20.34 sec	-
Cowell	75	14.89 sec	.2 D-6 km
Cowell	50	9.69 sec	.23 D-5 km
Cowell	40	7.44 sec	.358 D-3 km
Cowell	30	6.12 sec	.21 D-1 km
D.S.	100	46.37 sec	.1 D-6 km
D.S.	75	34.32 sec	.1 D-6 km
D.S.	50	23.24 sec	.2 D-6 km
D.S.	40	18.84 sec	.11 D-5 km
D.S.	30	14.89 sec	.14 D-4 km
D.S.	20	10.35 sec	.14 D-2 km
D.S.	15	10.80 sec	.70 D-2 km

Table 2

Satellite: AE-C Flight 68 days 12 hrs.
Forces: 4 x 4 Earth's Gravitational Field, Sun, Moon Drag
Semi-Major Axis at Final Time: 8384.257 km

Orbit Generator	Steps/rev.	CPU Time	Error in SMA
Regularized Cowell n = 1.5	70	2 min 4.42 sec	.055 km
D.S.	70	4 min 22 sec	.016 km
D.S.	40	2 min 51 sec	.028 km
D.S.	20	2 min 17 sec	.735 km

Table 3

Satellite: SMS Transfer Flight 140 days
Forces: 4 x 4 Earth's Gravitational Field, Sun, Moon
Semi-Major Axis at Final Time: 24455,75850059 km

Orbit Generator	Steps/rev.	CPU Time	Error in SMA
Regularized Cowell n = 1.5	200	2 min 15.34 sec	.31 D-6 km
Regularized Cowell n = 1.5	150	1 min 43.35 sec	.77 D-5 km
Regularized Cowell n = 1.5	120	1 min 21.87 sec	.174 D-3 km
Regularized Cowell n = 1.5	90	1 min 1.79 sec	.831 D-2 km
D.S.	100	2 min 26.77 sec	.42 D-6 km
D.S.	75	1 min 50.49 sec	.59 D-6 km
D.S.	50	1 min 15.74 sec	.28 D-4 km
D.S.	40	1 min. 1.05 sec	.66 D-3 km
D.S.	30	52.39 sec	.26 D-2 km

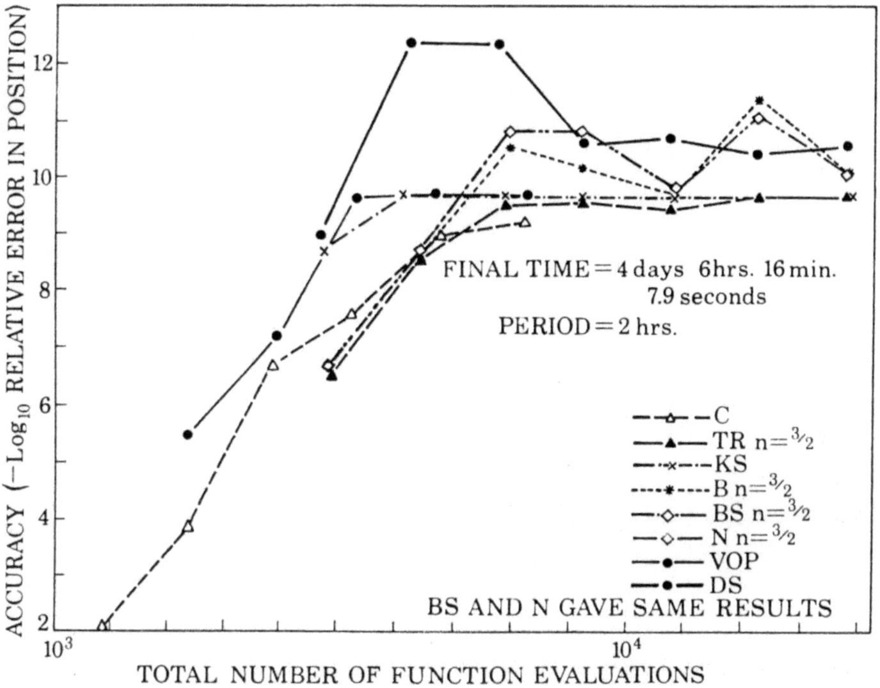

Figure 1. GEOS-B accuracy graphs.

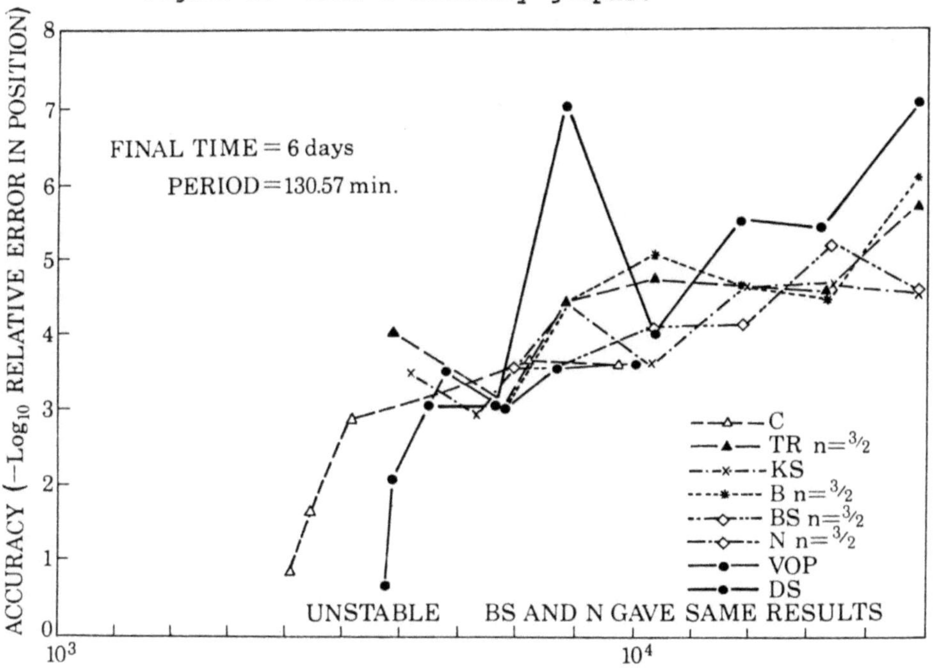

Figure 2. AE-C accuracy graphs.

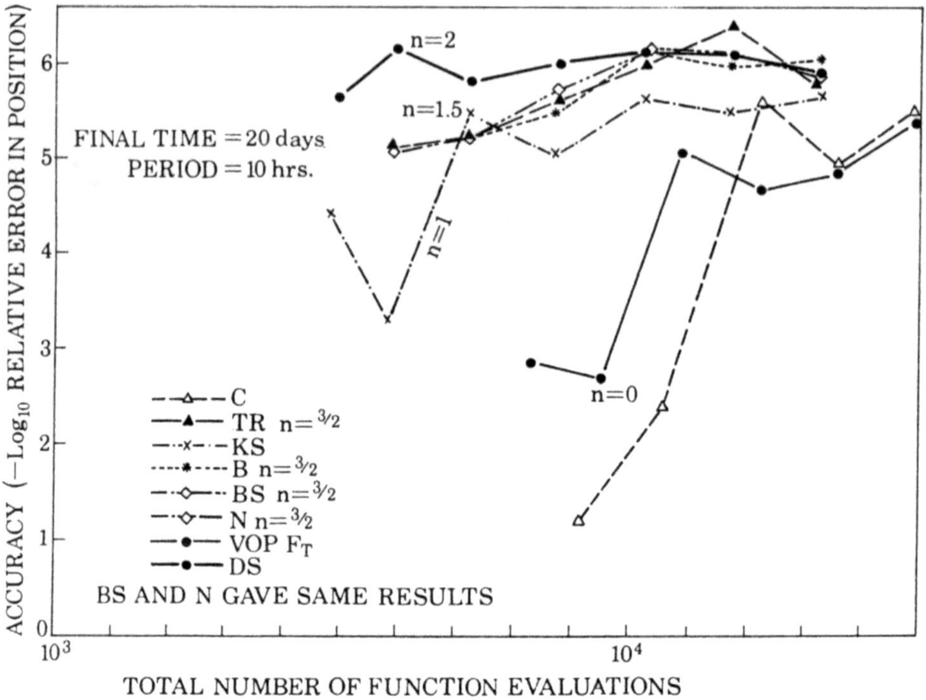

Figure 3. SMS accuracy graphs.

References

1 Velez, C.: Notions of Analytic vs. Numerical Stability as Applied to the Numerical Integration of Orbits. Presented at the AIAA/AAS Astrodynamics Conference, Vail, Colorado, July 1973, to appear in Celestial Mechanics.

2 Long, A.; Nimitz, K.; Cefola, P.: The Next Generation of Orbit Prediction Formulations for Artificial Satellites II. CSC, 9101-14600-01TR, March 1973.

3 Stiefel, E.; Scheifele, G.: Linear and Regular Celestial Mechanics. New York: Springer-Verlag 1971.

4 Baumgarte, J.: Dynamic Stabilization of Perturbed Keplerian Motions. Celestial Mechanics 5 (1972) 490.

5 Baumgarte, J.; Stiefel, E.: Celestial Mechanics (1974) in press.

6 Moore, W.; Beaudet, P.: The Testing of Fixed-Step Numerical Integration Processes for the Cowell Method of Special Perturbations. Proceeding of the Conference on the Numerical Solution of Ordinary Differential Equations, v. 362. Heidelberg: Springer-Verlag, p. 149.

7 Nacozy, P.: Time Elements. Presented at Proc. Symp. on Satellite Dynamics. Berlin: Springer-Verlag, 1974.

8 Beaudet, P.: The Testing and Comparison of Various Methods of Special Perturbations. CSC 3000-08600-01TM, March 1974.

9 Baumgarte, J.; Stiefel, E.: Examples of Transformations Improving the Numerical Accuracy of the Integration of Differential Equations. Procedures of the Conference on the Numerical Solution of Ordinary Differential Equations, Springer-Verlag, p. 207.

10 Nacozy, P.: University of Texas at Austin, private communication.

11 Scheifele, G.: Numerical Orbit Computation Based on a Canonical Intermediate Orbit. Report for ESRO, ESOC-Contract 490/72/AR, September 1973.

12 Samway, R.: A Special Perturbation Method Based on Canonical Delaunay-Similar Elements with the True Anomaly as the Independent Variable. AMRL 1054, University of Texas at Austin, July 1973.

13 Peñas, A.: Accurate Integration of Orbits Using Delaunay-Similar Elements. GSFC X-document 582-73-375, December 1973.

Orbit Determination in the Presence of Atmospheric Drag Errors *

B. D. TAPLEY[1], A. R. NETO[2] and B. E. SCHUTZ[3]
Department of Aerospace Engineering and Engineering Mechanics, The University of Texas at Austin 78712 USA

Abstract

A method for estimating the state of a spacecraft in the presence of unmodeled aerodynamic forces due to fluctuations in the atmospheric density and unknown rotational motion of the spacecraft is presented. The proposed method, based on a Dynamic Model Compensation approach, processes the data in a sequential fashion. In addition to estimates of the state, sequential estimates of the unmodeled aerodynamic accelerations are determined also. The estimates of the unmodeled accelerations can be used to improve knowledge of both the vehicle drag parameter and the atmospheric density.

Introduction

The determination and prediction of the orbit of a satellite in the near-earth environment is complicated by the fact that the satellite is influenced by the nonspherical effects of the earth's gravitational field as well as the dissipative effects of the earth's atmosphere. The effects of the atmosphere are difficult to determine because the atmospheric density, and hence, the drag undergo large unmodeled fluctuations. For many artificial satellites, this fluctuation in the drag is a fundamental source of error in the orbital predictions. The fluctuations in the drag can be correlated with the

---

* This research was supported by the Air Force Office of Scientific Research under Grant Number 72-2233.

[1] Professor and Chairman.

[2] Presently Assistant Professor of Mechanical Engineering, University of São Paulo, São Paulo, Brazil.

[3] Assistant Professor

dynamic state of the earth's atmosphere, i.e., the hourly, daily, and monthly fluctuations in the atmospheric density, as well as in the orientation of the spacecraft. This latter effect will directly influence both the drag coefficient and the projected cross-sectional area. In this investigation, a method for estimating the state of a spacecraft in the presence of such unmodeled aerodynamic forces is described. The proposed method has the advantage that, in addition to obtaining a more precise estimate of the state of the spacecraft, values of the unmodeled aerodynamic accelerations are determined also as a function of time. The estimates of the unmodeled aerodynamic accelerations can be used to improve knowledge of both the vehicle drag parameter and the atmospheric density. With this improved knowledge, the long-term prediction capability can be improved.

Mathematical Approach

The equations which describe the motion of a nonlinear dynamical system can be expressed as follows:

$$\dot{r} = v, \qquad \dot{v} = a_m(r,v,t) + m(t) \tag{1}$$

where r is a three vector of position components; v is a three vector of velocity components; and $a_m(r,v,t)$ is a three vector of modeled acceleration components. The three vector m(t) designates the unmodeled acceleration components, and it represents the effects of all the accelerations which are not accounted for in the mathematical model used to describe the physical process. Some of these effects such as those due to parametric uncertainties can be handled by treating certain parameters in the modeled accelerations in Eq. (1) as unknown, but constant, parameters and estimating their value during the orbit determination process. However, in situations where a large number of such parameters are required, as in the description of the earth or lunar gravitational fields, this approach will lead to computational difficulties. Other effects such as radiation pressure, translational forces due to unbalanced attitude jets, fluctuations in atmospheric drag, random error in propulsion systems, etc., depend on such a large number of parameters that they can be best approximated as a continuous random process.

In the Dynamic Model Compensation approach used in this investigation, the unmodeled acceleration m(t) is approximated by the function $\varepsilon(t)$ where the (3×1)-vector $\varepsilon(t)$ is assumed to be governed by differential equations which have the following general structure,

$$\dot{\varepsilon} = g(\varepsilon,\alpha,t) + u_\varepsilon \quad . \tag{2}$$

The vector α represents a set of unknown parameters which are adjusted to allow ε to give the best possible approximation to m(t). The three vector u_ε is assumed to be Gaussian noise whose statistics are

$$E\left(u_\varepsilon\right) = 0, \qquad E\left(u_\varepsilon(t)u_\varepsilon(\tau)\right) = q_\varepsilon\delta(t-\tau) \quad . \tag{3}$$

In Ref. 1-4, a first-order Gauss-Markov process, i.e., the Langevin equation, is used to approximate the unmodeled accelerations due to anomalies in the earth and lunar gravitational potentials. In Refs. 1 and 2, results obtained by processing range-rate tracking data during the lunar orbiting phase of the Apollo 10 and 11 missions indicate that the estimates of the unmodeled accelerations are repeatable from orbit to orbit within a given mission and from mission to mission when the same ground track is covered. In Ref. 4, it is shown that these estimates are an accurate approximation to the acceleration anomalies due to lunar surface mascons. In Ref. 5, the application of both first-order and second-order Gauss-Markov processes to the estimation of the random forces acting on a low-thrust space vehicle is described. In these studies, it is shown that, if a sufficiently accurate and sufficiently dense data set is available, the unmodeled accelerations may be estimated to a high degree of accuracy. In the investigation described here, the unmodeled accelerations are assumed to be due to (1) the effects of random variations in the earth's atmospheric density and (2) to the effects of variations in the drag parameter due to changes associated with the satellites rotational motion about the center of mass. The basic problem considered is that of estimating the state of a vehicle whose initial orbit as well as its initial vehicle parameters and initial rotational motion are unknown and, in the process, determining a precise estimate of the atmospheric density.

Model Compensation Approach

In the Model Compensation approach used in this study, it is assumed that the unmodeled accelerations are due primarily to atmospheric drag. With this assumption, the unmodeled accelerations are

$$m = B_c \rho |v_a| v_a + u_0 \tag{4}$$

where B_c is equal to the drag parameter divided by two, ρ is the atmospheric density, and v_a is the relative velocity of the space vehicle with respect to the atmosphere. The atmosphere is assumed to rotate at the same angular velocity as the earth. The three vector u_0 represents any additional random anomalies which influence the motion. Under the assumption that the atmospheric density can be approximated locally by an exponential function and the altitude and the drag parameter can be approximated locally as a sinusoidal variation about some mean value, the following approximations are made:

$$\rho = \rho_0 \exp(-\beta(r-r_0)), \quad B_c = B_0(1 + \varepsilon\cos(\omega t+\phi)) \tag{5}$$

where r_0 is a reference radius, ρ_0 is the density at radius r_0, B_0 is a mean drag parameter, and ε, ω, and ϕ are constants. Since it is of interest to obtain representations which will approximate the aerodynamic drag throughout the entire tracking period, the following differential equations are adopted for the description of ρ and B_c:

$$\begin{aligned} \dot{\rho} &= d(\rho_0 \exp -\beta(r-r_0))/dt + u_1 \\ \dot{\beta} &= u_2, \quad \dot{B}_0 = u_4, \quad \dot{\omega} = u_5 \\ B_c &= \omega^2 (B_c - B_0) + u_3 \quad . \end{aligned} \tag{6}$$

The random noise components u_i, $(i = 0,1,...,5)$, are each assumed to satisfy the conditions $E(u_i) = 0$ and $E(u_i(t)u_j(\tau)) = q_i(t)\delta(t-\tau)\delta_{ij}$ where $\delta(t-\tau)$ is the Dirac Delta function and δ_{ij} is the Kronecker Delta. If Eq. (1), with Eq. (3) substituted for m, is combined with Eqs. (6), the differential equations governing the model compensated motion can be expressed as follows:

$$\begin{aligned} \dot{r} &= v \\ \dot{v} &= a_m(r,v,t) + X_7 X_9 |v_a| v_a \\ \dot{X}_7 &= X_8 \\ \dot{X}_8 &= -X_{10}^2 (X_7 - X_{11}) + u_1 \\ \dot{X}_9 &= \frac{d}{dt} \{X_9(t_i) \exp(-X_{12}(r-r_i))\} + u_2 \\ \dot{X}_{10} &= u_3 \\ \dot{X}_{11} &= u_4 \\ \dot{X}_{12} &= u_5 \end{aligned} \tag{7}$$

where v_a is the velocity of the satellite with respect to the atmosphere. Note that r and v are 3-vectors and to complete the description of the equations of motion, a coordinate system must be selected and the first two of Eqs. (7) expressed in component form. In the investigation described here, a geocentric cartesian coordinate system was used where the x-y plane coincides with the earth's equator, the positive x-axis pointing towards the Vernal Equinox of 1950 and the positive z-axis coinciding with the earth's angular velocity vector. The differential equations governing the model compensated motion can be expressed then as

$$\dot{X} = F(X,t) + U \tag{8}$$

where the 12-vector X is composed of the six components of r and v plus the six components of the unmodeled acceleration. The 12-vector of nonlinear functions, F(X,t) is defined by the right hand sides of Eq. (7) if the u_i are omitted and

$$U^T = \left(0\ 0\ 0 : 0\ 0\ 0 : 0\ u_1\ u_2\ u_3\ u_4\ u_5\right) \quad . \tag{9}$$

Note that U satisfies the condition

$$E\left(U\right) = 0 \qquad E\left(U(t)U^T(\tau)\right) = Q(t)\ \delta(t-\tau) \quad . \tag{10}$$

The matrix Q is diagonal with $Q_{jj} = 0$, $i = 1,\ldots,7$, $Q_{jj} = q_{j-7}$; $i = 8,\ldots,12$ and $Q_{ij} = 0$ for all $i \neq j$.

It is assumed that the observations of the satellite satisfy the following observation-state relationship

$$Y_i = G(X_i,t_i) + \eta_i \quad , \quad i = 1,\ldots,m \tag{11}$$

where Y_i is a p-vector of observations at the observation epoch, t_i and n_i satisfies the condition $E\left(\eta_i\eta_i\right) = R_i$.

With the sequence of observations Y_i; $i = 1,\ldots,m$, the estimate of the state can be obtained by using the Kalman-Bucy filter (6) in its extended form. In the extended sequential filter, the trajectory is rectified at each observation point to include the current estimate. The prediction between observations is obtained by using the current estimate of the state and the state-error covariance matrix as the initial conditions for the propagation. The computational algorithm resulting from this procedure can be expressed in the following form:

$$\bar{X}_k = \hat{X}_{k-1} + \int_{t_{k-1}}^{t_k} F(\bar{X},\tau)\,d\tau$$

$$\bar{P}_k = P_{k-1} + \int_{t_{k-1}}^{t_k} \left(A(\tau)\bar{P} + \bar{P}\,A^T(\tau) + Q\right)d\tau$$

$$K_k = \bar{P}_k H_k^T \left(H_k \bar{P}_k H_k^T + R_k\right)^{-1}$$

$$\hat{X}_k = \bar{X}_k + K_k\left(Y_k - G(\bar{X}_k, t_k)\right)$$

$$P_k = \left(I - K_k H_k\right)\bar{P}_k \tag{12}$$

where $H_k = \left(\partial G/\partial X_k\right)^*$, $A = \left(\partial F/\partial X_k\right)^*$, and $()^*$ implies that the quantity in the brackets is evaluated on the solution, $X(t) = \bar{X}(t)$. See Ref. (7) for a more complete discussion of the extended sequential estimation algorithm.

Numerical Results

The characteristics of the algorithm described in the previous section have been evaluated through a study of the orbit determination problem for a near-earth satellite. The numerical results presented in this section were obtained through numerical simulation of the orbit determination procedure. In the simulated study, the density for the true trajectory is calculated using an analytic approximation to the 1970 Jacchia atmosphere (8). The gravitational potential is the 1969 Smithsonian Standard Earth II with the spherical harmonic coefficients truncated at degree and order four with J_5 and J_6 included (9). The lunar and solar perturbations are included assuming point mass and elliptic two-body motion for their ephemeris. In the estimation algorithm, the drag effects are approximated as unmodeled accelerations and the algorithm described in the previous section is applied in an attempt to estimate the variations in the drag force. There will be erroneous aerodynamic forces due to density errors associated with errors in the position of the spacecraft. In addition, it is assumed that $B_c = C_d S/m$ varies with the rotation rate and it is further assumed that C_d, S/m and the rotation rate are unknown. Values are adopted for these quantities in order to generate a true trajectory and the true trajectory is used then to generate true observations. The true observations are corrupted by the addition of a zero mean random noise and these corrupted observations are used by the estimation algorithm in an attempt to obtain an accurate estimate of the state of the vehicle along with estimates of the unmodeled drag accelerations. The vehicle whose trajectory is to be estimated is initially in a 250 km altitude circular orbit. The initial state of the vehicle is as follows: $x_0 = -4165.72$ km, $y_0 = -3949.43$ km, $z_0 = 3807.12$ km, $\dot{x}_0 = 6.4228$ km/sec, $\dot{y}_0 = -3.8175$ km/sec, and $\dot{z}_0 = 3.0173$ km/sec. The initial position of the space vehicle is just above the western coast of the United States. The orbital inclination is approximately 60^0. A sequence of 11 tracking stations is used to obtain range and range rate observations of the space vehicle as it orbits the earth. The locations of the tracking stations and the relative positions with respect to the satellite ground track are shown in Fig. 1. The observation accuracies assigned to the tracking stations are $\sigma_\rho = 5$ m and $\sigma_{\dot\rho} = 1\times10^{-3}$ m/sec. In addition to the errors in the dynamic model used for the motion of the satellite by the estimation procedure, the true trajectory and the initial reference trajectory differ by 150 m in each component of position. The state error covariance matrix is initially assumed to be diagonal with $P_{xx} = P_{yy} = P_{zz} = 100$ m and $P_{\dot x\dot x} = P_{\dot y\dot y} = P_{\dot z\dot z} = 100$ m/sec.

Figure 2 shows a comparison between the magnitude of the estimated error in position, Δr, compared with the trace of the covariance elements associated with the position error. Note that, after the initial transient period during which the estimation procedure is acquiring the orbit, the error in the estimate is never larger than 5 meters in position. A maximum error in velocity of 9 cm/sec occurs during this same interval. The largest error in Fig. 2 occurs around 85 minutes during a 15 minute interval in which no stations observe the vehicle. The number of stations tracking the vehicle during any given interval are shown at the bottom of the figure. Note further that during the dense tracking period around 19 minutes an error of the order of 5 cm in position and 5 mm/sec in velocity is obtained. Finally, after one complete revolution (at approximately 100 minutes) errors of the order of 3 cm in position and 8 mm/sec in velocity are achieved. The results for the velocity error and covariance matrix are similar to those shown in Fig. 2. Figure 3 shows a comparison of the x component of the unmodeled aerodynamic acceleration compared with the estimate of the acceleration. These quantities are designated m_x and ε_x, respectively. Note that the estimate of the unmodeled acceleration approximates the mean of the unmodeled aerodynamic acceleration throughout the trajectory. Around 45 minutes the estimate begins to achieve the proper frequency, but the amplitude is only one-half of the amplitude of the true aerodynamic acceleration. Following the observation interval which terminates at 60 minutes, the estimate is a good approximation in both amplitude and frequency. In Fig. 4, the state variable, X_7, which represents the drag parameter, is shown as a function of time. Note that during regions in which the vehicle is not being tracked a sinusoidal variation is predicted. This variation is associated with the rotation of the vehicle about its center of mass. Figure 5 shows the mean value of the drag parameter and the variation should correspond to the average of the results shown in Fig. 4. Note that while a positive value was assumed for the initial guess for X_8, the estimate assigns a negative value. The negative value is consistent with the fact that the aerodynamic force was assigned a positive value in the original equations of motion and, if the drag force is to have the correct sign, a negative value for X_8 must be obtained by the estimation algorithm. Finally, Fig. 6 shows a comparison of the true atmospheric density, ρ, and the estimate of the atmospheric density, i.e., X_9. Note that the variations between the estimate and the true density are of the order of $.02\times10^{-12}$ gm/cm^3. The periodic variation in the amplitude is associated with a slight eccentricity in the orbit. For all times after the epoch of 60 minutes, a quite accurate estimate of the atmospheric drag is achieved.

Conclusions

Based on the results presented in the previous section, it can be concluded that the model compensation method used to approximate the aerodynamic forces for a space vehicle whose drag parameter and rotational motion are unknown will allow a precise estimate of the state using currently operational tracking networks and current operational radar measurements. Furthermore, for sufficiently accurate observations, an accurate estimate of both the true atmospheric density and the vehicle drag parameter can be obtained.

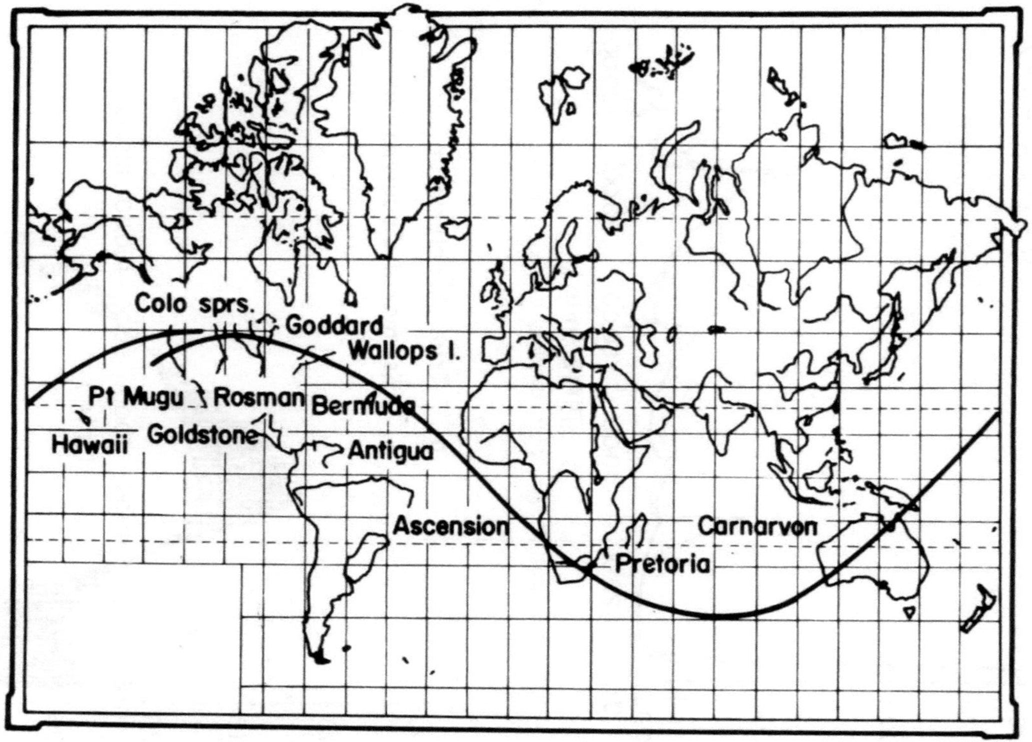

Figure 1. Tracking station location and satellite ground track.

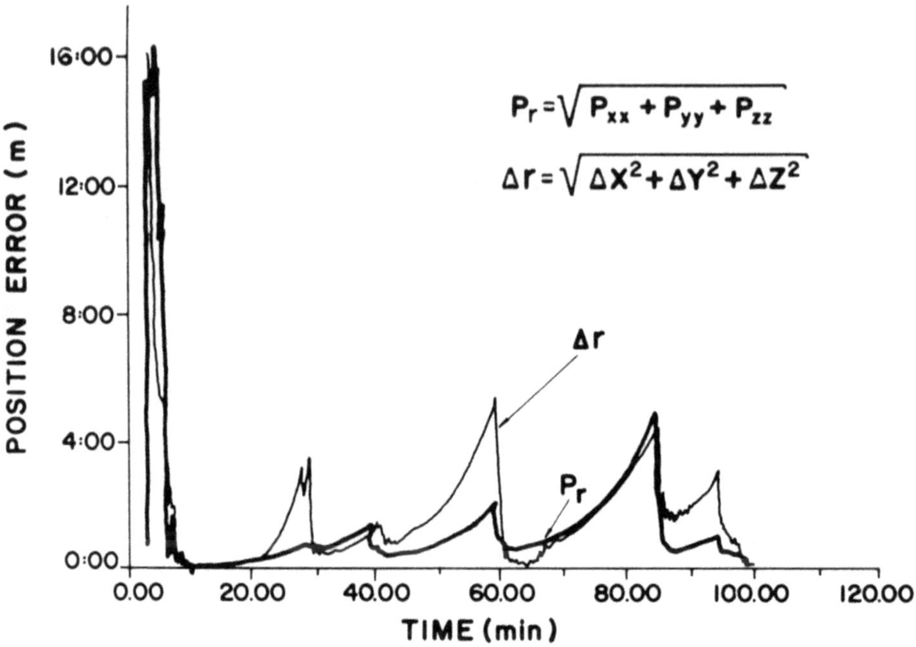

Figure 2. Position error versus time.

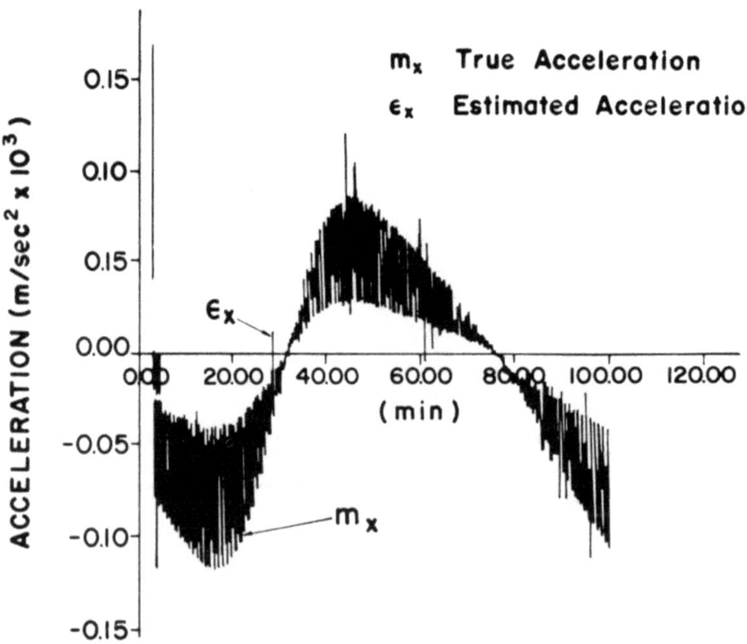

Figure 3. X-component of the unmodeled acceleration versus time.

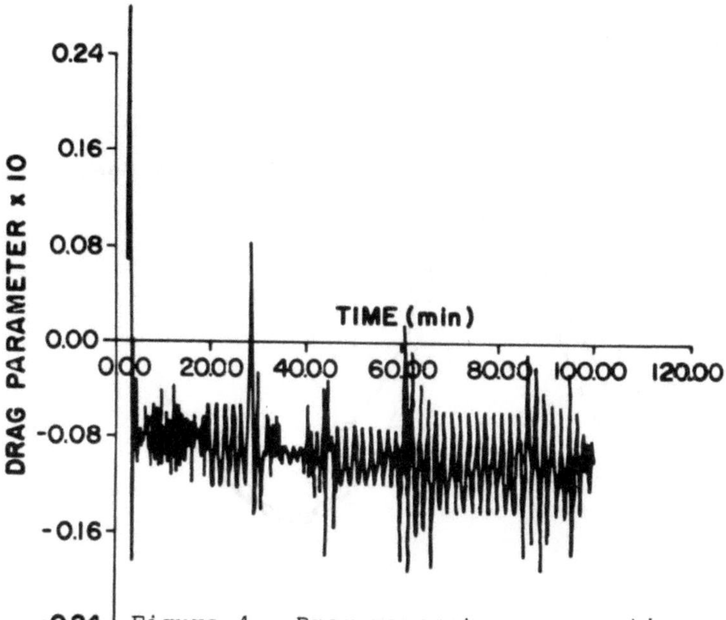

Figure 4. Drag parameter versus time.

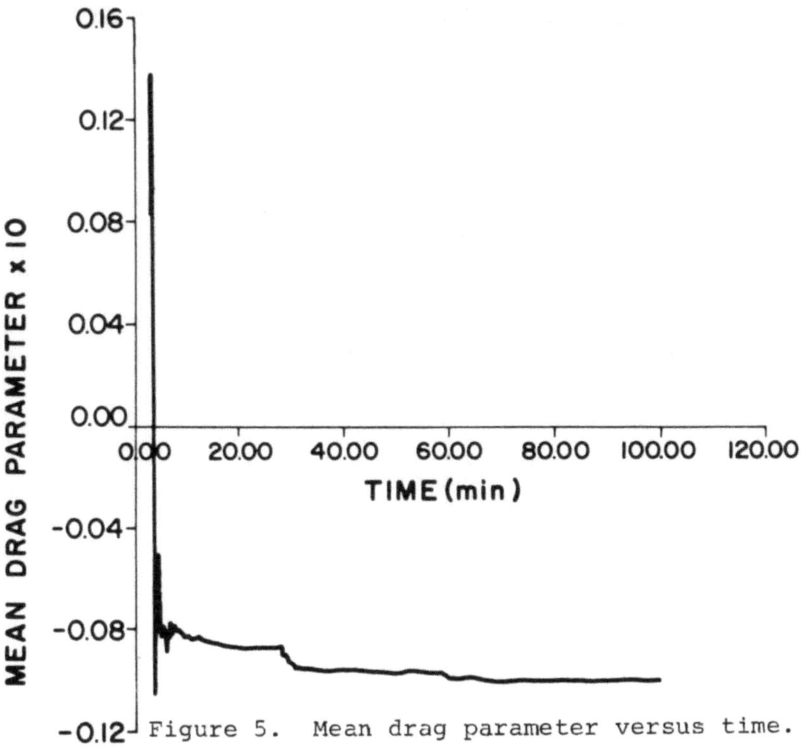

Figure 5. Mean drag parameter versus time.

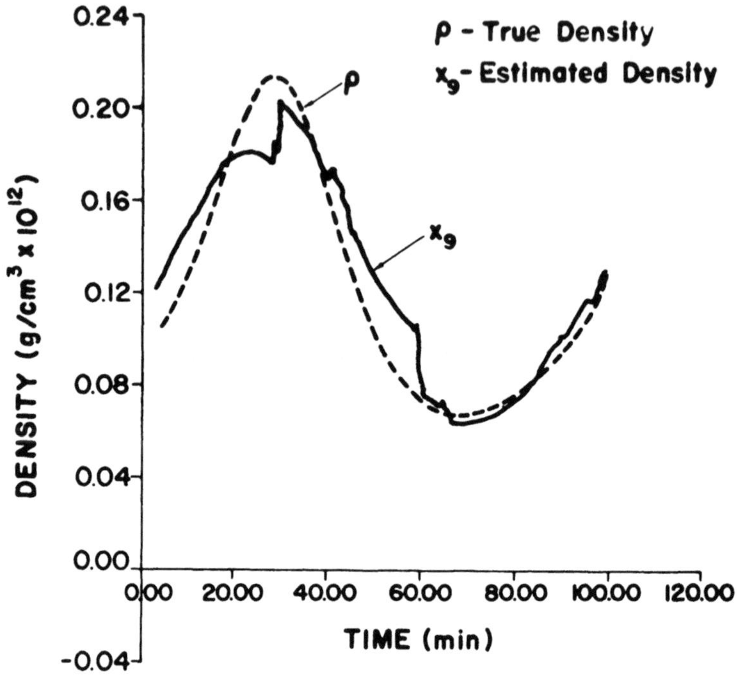

Figure 6. Atmospheric density versus time.

References

1 Tapley, B.D.; Ingram, D.S.: Orbit Determination in the Presence of Unmodeled Accelerations. Proc. of the 2nd Symp. on Nonlinear Estimation Theory. San Diego, September 1971.

2 Ingram, D.S.; Tapley, B.D.: Lunar Orbit Determination in the Presence of Unmodeled Accelerations. AAS/AIAA Astrodynamics Specialist Conference, Paper No. 71-371, Ft. Lauderdale, Florida, 1971.

3 Schutz, B.E.; Tapley, B.D.; Connolly, P.E.: Orbit Determination in the Presence of Gravity Model Errors. AGU Symp. on Earth Gravity Models, St. Louis, Missouri, August 1972.

4 Tapley, B.D.; Schutz, B.E.: Estimation of Unmodeled Forces on a Lunar Satellite. 23rd Cong. of the Int. Astr. Fed., Vienna, Austria, October 1972.

5 Tapley, B.D.; Hagar, H.: Estimation of Unmodeled Forces on a Low-Thrust Space Vehicle. Proc. of the 3rd Symp. on Nonlinear Estimation Theory. San Diego, September 1972.

6 Kalman, R.E.; Bucy, R.S.: New Results in Linear Filtering Prediction Theory. J. Basic Engr., Trans. ASME 83-D (1961).

7 Tapley, B.D.: Statistical Orbit Determination. Recent Advances in Dynamical Astronomy, ed. by B.D. Tapley and V. Szebehely. Dordrecht: D. Riedel Publishing Co. 1973.

8 Roberts, C.E., Jr.: An Analytic Model for Upper Atmospheric Densities Based Upon Jacchia's 1970 Models. Celestial Mechanics II (1971).

9 Gaposchkin, E.M.; Lambeck, K.: 1969 Smithsonian Standard Earth (II), Smithsonian Astrophysical Observatory. Special Report 315 (1969).

Literal Algebra for Satellite Dynamics

E. M. GAPOSCHKIN
Center for Astrophysics, Harvard College Observatory and Smithsonian Astrophysical Observatory, Cambridge, Massachusetts 02138 USA

Abstract

Analytical developments of a satellite perturbation theory can be accomplished quickly, accurately, and with generality by use of computer algebra. The basic principles of computer algebra, the required and desirable features of computer-algebra programs, and the applications of computer algebra to specific problems in perturbation analysis will be presented here. The emphasis of non-numeric mathematics in computer science, its implications for satellite theory, and the relation of the tool (computer algebra) to the problem (perturbation analysis) will be discussed. The use of computer algebra changes one's approach to perturbation theory, eliminating one class of problems (e.g., algebraic blunders) while introducing another (e.g., programing mistakes). Larger problems can be solved quickly; however, careful verification is still necessary. The development of computer science is oriented toward question solvability; uniqueness and large problems (efficiency) are not of interest. Conversely, perturbation theory benefits from the automated treatment of long series.

Introduction

Digital computers have been carrying out complex numerical calculations for a long time, opening possibilities for analysis inconceivable without them. The use of computers to perform formal algebraic and other non-numeric tasks has developed only in the last decade.

Computers are as useful for non-numeric application as they are for numerical computation, and one can question why computing has had the preponderance of numerical

---

This work was supported in part by grant NGR 09-015-002 from the National Aeronautics and Space Administration.

applications. Certainly, the tradition in analysis has been formal rather than numerical. Since computers are currently designed for numerical applications, they are awkward to use for non-numeric work. For example, computers normally have hardware to multiply numbers, but as yet, multiplication of algebraic expressions must be done with programs.

Nevertheless, non-numeric use of computers has grown considerably. Literal computer algebra has become relatively common and now takes its proper place as one of the tools available to the modern scientist. Algebra systems are not so systematized as, for example, FORTRAN, yet one can see the time shortly when rather general-purpose algebra systems will be routinely used in the same way as FORTRAN is for numeric applications. We hope that attention will soon be given to the problems being solved rather than to the tool. We do not devote much time to a description of FORTRAN or a fountain pen.

The plan of this paper is to describe the rather general class of operations available, to relate them to problems in satellite dynamics, to indicate some important characteristics of non-numeric calculation, and to point out some areas that can be resolved in the near future. However, the remarks reflect today's capability. One cannot strike a balance between efficiency and generality in a vacuum. Pointless generality misuses programming and design effort, as well as computer resources, and may make some problems inaccessible. Overemphasis on efficiency limits the scope of the system with the same result. The compromise is made in the context of the applications area-viz, celestial mechanics. Design of particular systems depends on the computer facility available. Implementation is discussed in papers such as those by Cherniack (3) and Davis (5).

Celestial mechanics, as is much of mathematical physics, is concerned with motions that are "nearly" periodic, and an approximation theory containing trigonometric functions results. Therefore, extraordinarily useful functions are polynomials and Poisson Series. This class of functions is simple to manipulate; and most important they are closed under the elementary operations of addition, subtraction, multiplication, and differentiation.

The first paper to deal with Poisson series (6) discussed the general term written as:

$$N\cdot 10^{n}\ x_1^a x_2^b x_3^c x_4^d x_5^e x_6^f x_7^g \ {\cos \atop \sin}\ (hA_1 + iA_2 + jA_3 + kA_4 + \ell A_5). \tag{1}$$

In Deprit's notation (5), this is a (7,5) system, i.e. there are seven polynomial variables and five trigonometric arguments. This has been generalized by Danby et al. (4) and is called Mechanized Algebraic Operations (MAO). A (10,7) system, MAO treats expressions

$$\sum_{i_1}\sum_{i_2}\cdots\sum_{i_m}\sum_{j_1}\sum_{j_2}\cdots\sum_{j_n} P_1^{i_1}P_2^{i_2}\cdots P_n^{i_m}\ C_{i_1 i_2\cdots i_m}^{j_1 j_2\cdots j_n}$$

$$\times\ {\cos \atop \sin}\ (j_1t_1 + j_2t_2 + \cdots + j_nt_n) \quad . \tag{2}$$

MAO was generalized to the Echeloned Series Processor (ESP) with the following general form:

$$\sum_i T_i \sum_j D_j \sum P_m \quad , \tag{3}$$

where

$$T_i = \frac{\cos}{\sin} (i_1 \ell_1 + i_2 \ell_2 + \cdots + i_6 \ell_6)$$

$$D_j = \frac{1}{\Pi_k (a_{jk} n_1 + \beta_{jk} n_2 + \gamma_{jk})^{**} \beta_{jk}}$$

$$P_m = \text{coefficient} \times P_1^{\gamma_1} P_2^{\gamma_2} \cdots P_{12}^{\gamma_{12}} \quad .$$

An alternative processor called TRIGMAN (7) started as a (9,10) system and has been extended to a (29,10) system. Finally, SPASM (3) allows general expressions of the form

$$\sum_i A_i \frac{\cos}{\sin} (B_i) \quad , \tag{4}$$

where A_i and B_i can be arbitrary expressions, generally polynomials. A number of other processors exist. The above-cited examples illustrate the degree of generality adopted by some users.

The implementation of an algebra processor depends on a) the design tradeoff of generality against efficiency and b) the available computer. These tradeoffs are beyond the scope of this article, and we confine ourselves to a few remarks. More details on implementation can be obtained from the references.

Algebra programs fall into two categories: a) table processors and b) symbol processors. Table-processing programs are very efficient but restricted. ALPAK, for example, is restricted to polynomials. Variables are associated with position in a table.

The four main categories of symbol processors follow, together with examples:

a) List processing: LISP1.5, IPL-V, SLIP;

b) String manipulation: COMIT, SNOBOL;

c) Symbol manipulation, a superset of b;

d) Formula manipulation: MATHLAB, FORMAC, ESP, TRIGMAN, SPASM.

There can be considerable overlap. For example, SPASM uses SLIP to manage lists and storage. TRIGRUN, a higher control language for TRIGMAN, is written in SNOBOL. SPASM uses a table processor when manipulating polynomials.

Fundamental Required Operations for an Algebra Processor

Closure: Clearly, any operation the algebra program performs must result in defined functions or expressions. But, for example in differentiation, we cannot allow for a function that does not have a derivative defined.

Addition: Self-evident.

Subtraction: Self-evident.

Multiplication: Self-evident.

Division: A great deal of work can be performed without division by expressions. A general problem in division concerns finding greatest common dividers, and it is now, still, unsolved. On the other hand, in the spirit of the successive approximations that pervade celestial mechanics, it is always possible to find a reciprocal of a function (2). Division by polynomials or monomials is considered essential for factoring common expressions or coefficients (see remarks on intermediate swell).

Differentiation: This is the key operation in most modern high-order methods that use canonical transformations.

Substitutions: Self-evident.

Input: Self-evident.

Output: Self-evident.

Desirable Features for an Algebra Processor

Rational fractions: Keeping rational fractions for coefficients will preserve exact expressions.

Uniqueness: Uniqueness implies that a canonical form exists in the processor; that is, there is one and only one representation of an expression.

Zero recognization: This depends on the availability of rational-fraction arithmetic and uniqueness. Clearly, for some problems, it is crucial to know whether the two expressions are identical. For example, to use the fundamental Lagrange or Poisson brackets as a test for a canonical transformation, the expression must be exactly zero.

Computed symbols: A processor must be able to use a computed number as a symbol in an expression.

Numerical evaluation: To compute the numerical value of an expression, we must have given numerical values for each of the symbols.

Punching of FORTRAN cards: the processor must be able to prepare the formulas derived by the algebra program, for use by a FORTRAN compiler, in such a way that transcription errors are eliminated.

General expressions: With a general algebra program, it is useful to manipulate any expressions that, for example, can be punched on the right-hand side of a FORTRAN arithmetic statement.

Definition of functions: In conjunction with the processing of general expressions, it is convenient for the algebra program to know that a symbol represents a function, which has a derivative or an integral, and to define that derivative in terms of other known functions.

Poisson Series: We have talked about Poisson Series as being a fundamental ingredient of algebra processors.

A great deal of celestial mechanics can be performed by treating sines and cosines without the structure of Poisson Series. Poisson Series have the added advantage of applying all known trigonometric identities automatically. Therefore, if an expression contains $\sin^2 + \cos^2$ and that expression is converted to a Poisson Series, which can be done unambiguously and economically, all simplifications are obtained. However, Poisson Series is not necessarily the most economical way to represent an expression involving trigonometric functions (see intermediate swell).

Bracketing: Significant simplification and shortening of expressions can be accomplished by factorization using brackets (see discussion on simplification and GCD algorithms).

Expansion: Manipulation of brackets with expressions now leads to the most economical form. Therefore, expansion of bracketed expressions is important.

Truncation: Truncation plays an important role in obtaining compact expressions, when formulas are developed in powers of a small parameter.

Traversing lists: The ability to operate on each term or subexpression in an expression is vital. For example, in the implementation of Von Zeipel's method for the satellite problem, it is necessary to separate the Hamiltonian into two parts, one part dependent on the mean anomaly of the satellite, and the other part, independent of it. This is done term by term, and two expressions are created.

Asking questions: In examining an expression or term, it is necessary to know the existence of a variable or the exponent of any particular variable.

Automatic storage management: It is characteristic of early algebra programs that the user was responsible for the details of the storage management (5).

Expanded character set: Current computer hardware and internal codes allow a wide variety of characters, yet the input/output media available today is restricted to the IBM Model 32 keypunch, which has a limited character set. Much more comprehensible results could be obtained by expanding the character set.

Interactive algebra: A characteristic of doing algebra is that the analyst experiments. One substitution or another, or a different form of the expression, is used to solve a problem, obtain some simplification, or perform a check. Often after only a few terms have been developed, it is quite evident whether the experiment has been a success. With batch processing, more computation is done than necessary. Interactive algebra would be an enormous benefit.

Noncommutative operations: A number of variables do not have convenient commutative characteristics - for example, matrices and operators. An algebra program that would allow definition of this commutative property would be most valuable in solving problems involving matrices.

Other Considerations

We have discussed the general characteristics desired in modern algebra programs. Existing programs already incorporate some or all of these features. We expect to find more common availability of these features as the user community grows and comes to use algebra programs as a tool.

Simplification and GCD algorithms: A general simplification problem is very difficult, and the practitioners of computer science are not very sanguine about it (1). For problems in celestial mechanics, however, some success can be enjoyed with rather ad hoc approaches. In general, even the problem of factoring polynomials in several variables is very difficult. But if we assume that polynomials can be factored into a product of monomials, then there is a deterministic process for finding each monomial. Having done that, the resulting product can be expanded and subtracted from the original expression to see if they are equivalent, assuming the arithmetic was done with rational fractions. Experience on a number of problems has indicated that this is a successful approach and that all simplifications are found in this way that would have been obtained by hand.

Intermediate swell: One important difference between numeric and algebraic operations on a computer is the noncommutativity of operations with respect to time and storage. Consider the following expression: $(\cos^{30}x)^{20} - (\cos^{20}x)^{30}$. As a trigonometric polynomial, this expression takes very little computer storage, and the operation is trivial. As a Poisson Series, we have to develop two 50-term Poisson Series and subtract them in order to end up with zero. In general, the only way to combat intermediate swell is to apply as many simplifications to the expressions along the way as we can. Alternatively, use of secondary storage is possible. This is particularly true with Poisson Series, as they are so highly structured (3).

Conclusions

a) Algebra programs have been used for a number of problems in celestial mechanics with great success.

b) The advantage of computer algebra is its accuracy and speed.

c) Standardization of algebra facilities is in progress.

References

1 Barton, D.; Fitch, J.P.: Applications of algebraic manipulation program in physics. Rep. Prog. Phys. 35 (1972) 235-314.

2 Broucke, R.: Construction of rational and negative powers of a formal series. Comm. ACM 14 (1971) 32-35.

3 Cherniack, J.R.: A more general system for Poisson Series manipulation. Celestial Mech. 7 (1973) 107-121.

4 Danby, J.M.A.; Deprit, A.; Rom, A.: The symbolic manipulation of Poisson Series. Math Note No. 432, Document D1-82-0481, Boeing Sci. Res. Lab. 1965.

5 Davis, M.S.: Review of non-numerical uses of computers. In "Recent Advances in Dynamical Astronomy. Ed. by B.D. Tapley and V. Szebehely. Dordrecht-Holland: D. Reidel Pub. Co. 1973, pp. 351-391.

6 Herget, P.; Musen, P.: Astron. J. 64 (1959) 11-20.

7 Jeffreys, W.H.: Automated algebraic manipulation in celestial mechanics. Comm. ACM 14 (1971) 538-541.

A Comparative Study of the Calculation of Partial Derivatives in Space Dynamics

H. G. WALTER
Astronomisches Rechen-Institut
Heidelberg, Fed. Rep. of Germany

Abstract

In the rigorous treatment of the partial derivatives of the instantaneous state vector of a spacecraft with respect to the initial state vector the pertinent variational equations are usually integrated numerically. Since the procedure is a laborious one, its optimization is desirable. It is to be expected that a broader knowledge of the mathematical aspects of the variational equations is remunerated by an improved computational efficiency.

The means for making the integration of the equations faster and more accurate in applications to orbit determination and error analysis are elaborated and compared in relation to the choice of coordinates and independent variables, and the types of spacecraft orbits. Starting with the classical form of the variational equations and the utilization of regularized coordin-Baker method of solving a system of linear differential equation and the utilization of regularized coordinates for the calculus of partial derivatives are investigated. Several case studies lead to distinctive results in terms of time consumption and accuracy.

Introduction

While orbit computation of natural and artificial celestial bodies has regularly attracted the mathematicians and has become the driving force behind the development of sophisticated numerical and analytical theories, only sparse attention has been given to the calculation of the concomitant partial derivatives. They play, however, an important role in the determination of orbits, in error estimation, navigation and trajectory optimization. Whenever the state transition or Jacobi matrix is wanted, its components, the partial derivatives, must be procured as a function of time.

Essentially, there are two conceivable approaches to obtaining the partial derivatives:

1. The analytical formulation while mathematically and computationally the most efficient means of calculating partial derivatives, encounters the severe limitation that the various perturbing accelerations usually do not lend themselves to an analytic solution analogous to the one established by Goodyear (1) for the unperturbed two-body motion. A simplification of the force model with the objective of attaining analytic formulae entails unpredictable effects which impair the accuracy of the results (Rice (2)).

2. The direct numerical integration of the variational equations provides a mathematically rigorous solution and is readily applicable in the presence of an open-ended variety of disturbing forces. Furthermore, it includes the possibility of dealing with parameters in addition to the components of the state vector such as physical constants, by simply extending the system of variational equations. This method, however, necessitates the simultaneous integration of the equations of motion and their variational equations and usually requires high order formulae and small integration steps, thus making the procedure time consuming.

The observations above gave rise to a comparative study of the efficiency of several procedures leading to the components of the transition matrix by numerical integration of the variational equations. Hereby emphasis is placed on their mathematical treatment in terms of regularized coordinates, choice of independent variables and factorization of state transition partials, rather than on the expediency of any of the customary integration techniques.

First of all, by applying the Newtonian method, the partial derivatives were calculated by straight integration of the variational equations in cartesian coordinates as a reference (3). Secondly, the Peano-Baker method tailored to orbital motion (4) was tried out. Its outstanding feature resembles the principle of perturbation theory as it provides correcting terms to the Jacobi matrix of the Keplerian motion. Finally, the KS-theory (5) was applied to study the effects of linearization on the variational equations after the introduction of a regularizing independent variable.

In each of the cases a circular and a highly eccentric orbit of an artificial earth satellite have been subjected to sample calculations in order to assess the sensitivity of the three procedures to the orbit type in addition to their general mathematical properties.

Notations and Mathematical Formulations

Any point in the physical space X^3 and in the four dimensional parametric space U^4 is denoted by $\underline{x}$ or $\underline{u}$, respectively. The equations of motion of a body in the gravitational field subject to the acceleration of a homogeneous spherical mass and additional small perturbing accelerations are given by

$$\ddot{\underline{x}} = \underline{f}(x,\dot{x},t) \quad , \tag{1}$$

where

$$\underline{f} = -k^2 \frac{\underline{x}}{r^3} - \frac{\partial V}{\partial \underline{x}} + \underline{P} \quad ,$$

$$r = |\underline{x}| \quad ,$$

V is the disturbing potential due to conservative forces which depend only on the position vector

$\underline{P}$ stands for the remaining perturbing acclerations.

With reference to Stiefel and Scheifele (5) the equations of motion in the parametric space become

$$\underline{u}'' + \frac{1}{\mu^2}\,\underline{u} = L^T(\underline{u})\,\underline{G} \quad , \tag{2}$$

where

$L^T(\underline{u})$ is the transpose of the KS-matrix,

$$\underline{G} = \frac{1}{2}\left(r\underline{P} - \frac{\partial(rV)}{\partial \underline{x}}\right) \quad ,$$

$$\mu^2 = \frac{2}{h} \quad , \quad h = \frac{k^2}{r} - \frac{|\dot{\underline{x}}|^2}{2} - V(\underline{x})$$

Variational Equations in Cartesian Coordinates

Using the customary abbreviations for Jacobi matrices let

$$D = \frac{\partial(\underline{x})}{\partial(\underline{x}_0, \dot{\underline{x}}_0)} \quad , \quad C = \frac{\partial(\underline{f})}{\partial(\underline{x}, \dot{\underline{x}})}$$

Then the system of variational equations follows from proper differentiation of Eq. (1) yielding

$$\ddot{D} = C(t)\begin{pmatrix} D \\ \dot{D} \end{pmatrix} \tag{3}$$

D is a 3 by 6 matrix and furnishes together with $\dot{D}$ the 36 partial derivatives of the instantaneous state vector with respect to the initial state vector.

Peano-Baker Approach to Variational Equations

The Jacobi matrix calculation by means of the Peano-Baker method is divided up into two parts: First of all the Jacobi matrix of the unperturbed motion is computed from closed analytical expressions (e.g., Sconzo (6)) and, secondly, a system of ordinary differential equations for the terms correcting for the perturbing accelerations is set up and subjected to the Peano-Baker integration (4). The combination of the two steps leads to the state transition matrix of the perturbed motion.

Variational Equations in the Parametric Space

Following the results of ESOC (7) the system of variational equations in the parametric space with the eccentric anomaly E as regularizing variable ensues from Eq. (2) and in principle is expressed by

$$\Xi'(E) = C(E)\ \Xi\ (E) \quad , \tag{4}$$

where Ξ is a 10 by 10 matrix and C is derived from the equations of motion in the U^4 space in analogy to C(t) of the physical space.

Numerical Comparisons

Following the delineations of Variational Equations in Cartesian Coordinates and Peano-Baker Approach to Variational Equations the variational equations on the one hand and the correcting terms on the other have been integrated over one revolution, thus allowing a direct comparison of the partial derivatives. Results are given in Tables 1 to 3 for an unperturbed highly eccentric orbit, for a perturbed highly eccentric orbit exposed to luni-solar and earth oblateness perturbations, and for a circular orbit exposed to earth oblateness and atmospheric drag perturbations. The eccentric orbit is described by the semi-major axis a = 113400 km and the eccentricity e = 0.94, and the circular orbit by r = 6860 km.

It suffices to confine the comparison to one partial derivative, e.g., $\partial x/\partial x_0$, since it turns out to be representative for all components of the state transition matrix. Integration time is quoted in relative figures, the unity being assigned to the integration of the variational equations in cartesian coordinates. The notations A, I, II and III refer to the closed analytical expressions, the Newtonian, the Peano-Baker and the KS method, respectively. In the cases of perturbed motion the solution of the variational equations in cartesian coordinates are considered as reference for accuracy assessment.

In the notation of ESOC (7) the association of the state transition matrix in the physical space to the one in the parametric space is expressed by

$$\frac{\partial(\underline{x},\dot{\underline{x}})}{\partial(\underline{x}_0,\dot{\underline{x}}_0)} = T(E)\;\Xi\;(E)T_0^{-1} \quad . \tag{5}$$

As the relation between the independent variables, time t and eccentric anomaly E of the two spaces is rather involved and since this impairs a direct comparison of the partial derivatives, the utility of the KS method is assessed by applying the following procedure:

1. Integration of a reference orbit, of the corresponding partial derivatives and of a suitable variational orbit by means of the Newtonian and the KS method.
2. Deduction of the variational orbit from reference orbit and corresponding partial derivatives in accordance with both methods.
3. Comparison of the variational orbits thus obtained in terms of accuracy, step sizes and computer time.

Results are shown in Tables 4 and 5 for a highly eccentric and a planar circular orbit. The typical orbit parameters are a = 136000 km, e = 0.95, and r = 6600 km, respectively. To express the relative computing time the KS method has been chosen as reference.

The coincidence in Table 4 of the numerically integrated coordinates on the one hand and those obtained from the reference orbit and the concomitant partial derivatives on the other is explained by the exactitude of the latter method when perturbations are absent.

Despite the larger computing time needed by the KS method for a circular orbit (Table 5), the KS variations are superior to the Newtonian ones because of the high degree of accuracy which can be ascribed to the quasi linearity of the variational equations in the parametric space. This superiority is still more salient for eccentric orbits for which, besides, the automatic step size control of the KS method takes effect.

Another more direct approach to the comparison of the KS and Newtonian methods for solving the variational equations deals with the computation of the partial derivatives in the U^4 and X^3 spaces using step sizes of magnitudes lessening until the values of the derivatives become stationary. Even for a circular orbit the KS method is distinctly superior to the Newtonian method by a factor of two in step number and computing time, and by a factor of 10 for a highly eccentric orbit.

Discussion

While the integration of the variational equations in cartesian coordinates is the most straightforward approach to the calculation of partial derivatives, it is also the most time consuming one. It is, therefore, not recommendable, when the perturbing accelerations are small enough to provide conditions under which the Peano-Baker and the KS methods can develop their favourable properties. This means for the Peano-Baker method that the extra requirements for factorization are remunerated by greater step sizes which will compensate for the augmented computing time. Likewise, the KS method profits in the case of small perturbations from the quasi linearity of the equations of motion in the parametric space and, as a consequence, the integration of variational orbits is rendered superfluous most of the time.

When, however, the perturbations exceed a certain size and, in particular, when they become irregular as in the presence of atmospheric drag or finite thrust, the distinctions between the three methods described here begin to vanish.

Acknowledgements

In performing the numerical comparisons the author was able to avail himself of the algorithms and test calculations provided by Messerschmitt-Bölkow-Blohm (MBB), Munich, Germany and Büro für Angewandte Mathematik - Dr. Haller - Zürich, Switzerland. The major part of this work was accomplished during the author's affiliation with ESRO.

Table 1. Differences of partial derivatives of an unperturbed highly eccentric orbit calculated by means of the closed analytical expressions (A), the Newtonian method (I) and the Peano-Baker method (II).

Number of Steps Regul. Step Size (sec/m)	$\left(\frac{\partial x}{\partial x_0}\right)_I - \left(\frac{\partial x}{\partial x_0}\right)_A$	$\left(\frac{\partial x}{\partial x_0}\right)_{II} - \left(\frac{\partial x}{\partial x_0}\right)_A$
206 4.28×10^{-6}	0.11×10^{-4}	-0.17×10^{-5}
411 2.14×10^{-6}	0.78×10^{-6}	-0.12×10^{-6}
821 1.07×10^{-6}	0.40×10^{-7}	-0.10×10^{-7}

Table 2. Partial derivatives of a perturbed highly eccentric orbit calculated by means of the Newtonian method (I) and the Peano-Baker method (II).

Numbers of Steps Regul. Step Size (sec/m)	$\left(\frac{\partial x}{\partial x_0}\right)_I \times 10^{-2}$	$\left(\frac{\partial x}{\partial x_0}\right)_{II} \times 10^{-2}$	Relat. Integ. Time
206 4.28×10^{-6}	0.2679333803	0.2679332515	1.8
411 2.14×10^{-6}	0.2679332750	0.2679332670	1.8
822 1.07×10^{-6}	0.2679332686	0.2679332681	1.8
1643 0.54×10^{-6}	0.2679332682	-	-

Table 3. Partial derivatives of a perturbed circular orbit calculated by means of the Newtonian method (I) and the Peano-Baker method (II).

Number of Steps Regul. Step Size (sec/m)	$\left(\frac{\partial x}{\partial x_0}\right)_I$	$\left(\frac{\partial x}{\partial x_0}\right)_{II}$	Relat. Integ. Time
206 4.11×10^{-6}	-0.4622853080	-0.4622849263	1.7
400 2.05×10^{-6}	-0.4622849397	-0.4622849250	1.7
800 1.03×10^{-6}	-0.4622849256	-	-
1600 0.51×10^{-6}	-0.4622849250	-	-

Table 4. Highly eccentric variational orbit of unperturbed motion deduced from a reference orbit and the respective partial derivatives.

Integrat. over 1 Revol.	x_1 (km)	x_2 (km)	x_3 (km)	Steps per Revolution	Relative Comp. Time
Numeric. Integr. of Variat. Orbit, Δx_2 = 0.1 km	0.0	-5888.8727	-3400.0000	200 (KS) 2000 (Newt.)	-
KS Variations	0.0	-5888.8727	-3400.0000	200	1.0
Newtonian Variations	-210.4085	-6349.6660	-3666.0341	2000	3.9

Table 5. Circular variational orbit under the influence of oblateness perturbations, which is deduced from a reference orbit and the respective partial derivatives.

Integrat. over 1 Revol.	x (km)	x_2 (km)	Steps per Revolution	Relative Comp. Time
Numeric Integr. of Variat. Orbit, Δx_1 = 10km	6609.8730	41.0090	40 (KS) 80 (Newt.)	-
KS Variations	6609.8730	41.0079	40	-
Newtonian Variations	6612.6787	41.5242	80	-
Integr. over 5 Revol.				
Numeric Integr. of Variat. Orbit, Δx_1 = 10km	6606.8259	205.0134	40 80	-
KS Variations	6606.8262	205.0083	40	1.0
Newtonian Variations	6676.3689	216.2372	80	0.7

References

1 Goodyear, W.H.: Completely General Closed-Form Solution for Coordinates and Partial Derivatives of the Two-Body Problem. Astron. J. 70 (1965) 189-192.

2 Rice, D.R.: An Investigation into the Effects of Using Simplified Force Models in the Computation of State Transition Matrices, AIAA Paper No. 67-123 (1967) 1-7.

3 Walter, H.G.: The Importance of Partial Derivatives in Space Dynamics, in Proceedings of the 8th ESRO Summer School. ESRO SP-68 (1972) 129-142.

4 Ditto, F.H.: Partial Derivatives Used in Trajectory Estimation. Celest. Mech. 1 (1969) 130-140.

5 Stiefel, E.L.; Scheifele, G.: Linear and Regular Celestial Mechanics, Berlin: Springer 1971.

6 Sconzo, P.: Explicit Expressions for the 36 Terms of a Jacobian Matrix Used in Orbit Computations. Mem. della Soc. Astr. Ital. 34 (1963) 3-18.

7 ESOC: The Utilization of Regularized Coordinates for the Differential Correction of Spacecraft Trajectories, prep. by J. Haller et al., Contract No. 419/72/AR-ESRO, 1973.

Resonance Phenomena at Rotations of Artificial and Natural Celestial Bodies

V. V. BELETSKII
Moscow

Abstract

In recent years the study of the rotational motions of celestial bodies has been concentrated upon the resonance (synchronization) of these motions. Synchronous rotations of the natural celestial bodies have been developed in the systems of the Jupiter, Saturn and Mars satellites. Such synchronizations have been also discovered in the rotation of Mercury and Venus. To these should be added the well-known synchronous rotation of the Moon. These facts allow us to consider the tendency to synchronous rotation as an objective Law of Nature. The passive stabilization systems of the artificial satellites make use of this tendency.

The resonance rotations of the celestial bodies are described by specifically constructed theories. These theories acknowledge synchronization of the rotational motion as a regular law of nature. The theory enables us to substantiate the empirical laws of the Moon's rotation, i.e, Cassini's laws, and to construct the generalized Cassini's laws to which Mercury's rotation is subjected.

The generalized Cassini's laws describe a double synchronization first, between the axial rotation of a celestial body and its orbital motion and second, between the motion of an axis of rotation of the body and the disturbed orbital procession.

The Moon has double synchronization 1:1 and 1:1, Mercury has also double synchronization 3:2 and 1:1.

This double synchronization described by the generalized Cassini's law can be successfully used for passive satellite stabilization, for example, for the double orbital stabilization of orbital stations.

Magnetic and gravitational field interaction creates the conditions for specific satellite stabilization; for example, synchronization of type 2:1 can be expected for some natural satellites of the great planets. To intensify magnetic stabilization the gravitational field can be used for the artificial satellites.

Resonance between the rotational and translational movements of a body leads to qualitatively new effects. At the expense of this resonance it is possible to change essentially the satellite orbit (in theory even from circular to parabolic orbit). Due to the smallness of interaction of the rotational and translational movements this fact is of purely theoretical interest; nevertheless it is discussed in the literature. The idea of "gravicraft" is based on this principle.

The present review describes briefly the history of development and investigation of resonance phenomena in the rotations of the natural and artificial celestial bodies. Here we use the facts and results of papers (1-3,39,40) and some other investigations.

Introduction

Resonance in the Axial Planet and Satellite Rotations in the Solar System

In recent years the resonance effects in the solar system have been widely discussed; in particular the resonances in the orbital motions (4) are taken into consideration.

The special role of resonances in the solar system is also seen in the rotational motion of the planets around their axes. Everybody knows that one side of the Moon faces the Earth - it is a brilliant example of 1:1 resonance between the periods of orbiting and the period of revolution. Optical observations of recent years show (3,5) that a number of the natural satellites of the planets in the solar system rotate like the Moon with the same 1:1 resonance. These are Jupiter's satellites Io, Europe, Kallisto; and Saturn's satellites Japet, Rhea, Tehys, Dione. Quite recently (6) it has been found from experiments carried out from the American interplanetary station "Mariner-9" the Mars' satellites Phobos and Deimos also rotate with 1:1 resonance when they face Mars always with the same side. The scheme of movement with 1:1 resonance is illustrated in Fig. 1.

For a long time it was considered that the period of Mercury's rotation about the axis coincided with that of its rotation about the Sun, i.e. Mercury like the Moon rotates with 1:1 resonance. In that case one side of Mercury must always face the Sun. In this way in 1889 Schiaparelli interpreted many of his observations for Mercury. The period of Mercury's rotation about the Sun is known to be 88 days; Schiaparelli concluded that the period of Mercury's axial rotation was also 88 days. For almost a century (till 1965) the optical observations of the different astronomers have "confirmed" Schiaparelli's results.

However, the development of new techniques (including space engineering) has made it possible to organize fundamentally new methods for planetary observation, i.e. radar observations. In 1965 radar observations of Mercury allowed Pettengill, Dyce and Shapiro (7,8) to place a period of Mercury's axial rotation at 59±3 days instead of 88. Dj. Colombo (9) noted that this estimated radar period of Mercury's rotation was about 2/3 of its (88 day) period of revolution, i.e. the relation of the angular velocity of rotation to the angular (average) velocity of revolution will be equal to 3:2. Then the data of Schiaparelli and his followers were reconsidered; it was shown that the data allowed different interpretations and that the period of rotation in 59 days was in keeping with these observations. Thus the resonance 3:2 for Mercury's rotation was discovered. The scheme of such resonance motion is shown in Figure 2.

The resonance for Venus rotation has phenomenal character. The history of its discovery is less dramatic than that of Mercury, but the results are interesting. Due to the thickness of its atmosphere observations of Venus can give, in principle, only measurements of the period of rotation of the upper atmospheric layers. It was found that Venus' atmosphere has a retrograde rotation (relative to the orbital motion); but only discrepant and unreliable estimates were obtained (to which the author of the present report contributed as a schoolboy (10)).

The position changed in 1962 as a result of radar measurements carried out in the USSR (11,12) and in the USA (13,14). It was found that Venus rotates around its axis in a direction opposite to that of the orbital motion and with the period of about 250 days and the inclination of its axis to the normal of the orbital plane 1.2º According to these refined data (13,14) the period of Venus' rotation is found to be equal to 243.23±0.1 days which agrees with the value of 243.16 days within the limits of accuracy of the measurements. This value indicates that Venus must face the Earth always on the same side in every low conjunction with the Earth. Goldreich and Peale (15) were, evidently, the first to appreciate it. Thus, at Venus' rotation there exists a resonance which is phenomenal and is related not only with Venus' orbital motion but with the orbital motion of the Earth.

Actually the period of the Venus-Earth system is τ = 583.92 days. During this time the Earth in its proper orbital motion (see scheme Fig. 3) describes an arc

$$\alpha_3 \approx 2\pi + 2\pi\cdot 06 = \omega_3 \tau \quad , \qquad (1)$$

where ω_3 is the Earth's mean diurnal motion. Analogously Venus in its orbital motion describes an arc

$$\alpha_B \simeq 2\pi \cdot 2 + 2\pi \cdot 0.6 = \omega_B \tau \tag{2}$$

rotating around its proper axis in the opposite direction (i.e. in the direction which is opposite to the orbital motion). Having the angular velocity Ω, Venus rotates around its axis in time τ in an arc

$$\beta = -2\pi \cdot 3 + 2\pi \cdot 0.6 = -\Omega\tau \quad . \tag{3}$$

In this manner in time τ all three angles α_3, α_B, β will have the same values $\approx 2\pi \cdot 0.6$ (disregarding multiples of 2π). Subtracting this last equation from the two previous equations one finds that

$$5 \cdot \frac{2\pi}{\tau} = \omega_B + \Omega$$

$$4 \cdot \frac{2\pi}{\tau} = \omega_3 + \Omega \quad . \tag{4}$$

This means that in time τ Venus makes 5 total revolutions relative to its own orbital reference frame, and relative to the Sun-Earth line it makes 4 total revolutions. From these two last resonance relations we obtain

$$\Omega = 4\omega_B - 5\omega_3 \quad . \tag{5}$$

The data on resonance characteristics of rotation of the Moon, Mercury and Venus are given in Table 2.

The wide distribution of the resonance motions supports the view of non-randomness of these phenomena; there must exist forces contributing to the appearance of commensurability in the motion.

A theoretical explanation of resonance motions must consider: a) stabilizing the conditions which support the celestial body steadily within the resonance regime; b) dissipative forces which bring the arbitrary initial body rotation within the influence of the resonance regime of motion.

The gravitational torques interacting between the body and other celestial bodies act as stabilizing factors; tidal phenomena in the solar system are among the dissipative forces considered.

On the Construction of a Theory for Stationary Resonance Rotation

Let us describe successively constructions leading to the creation of a theory for the resonance rotational motions of celestial bodies. We shall start from the best known and most investigated case, the Moon's rotation.

In 1963 Cassini discovered the laws of the Moon's rotation (16,17):

1) The Moon rotates uniformly about an axis which remains fixed in the Moon's body, the period of this rotation coinciding with the period of its revolution in the orbit about the Earth.

2) The plane of the Moon's equator has a constant inclination to the ecliptic equal to 1°32.1.

3) The ascending node of the Moon's equator on the ecliptic always coincides with the descending node of the Moon's orbit on the ecliptic.

Figure 4 gives the scheme illustrating Cassini's law. As the Moon's orbit is inclined to the ecliptic at an angle of about $\tau \approx 5°09'$ it follows from Cassini's second and third laws that the axial angular velocity vector $\vec{\omega}$ of the Moon is normal to the line of the nodes and forms an angle $\rho \approx 6^\circ 41'$ with the normal $\vec{\eta}$ to the plane of the Moon orbit.

Starting from Lagrange and Laplace the theory of the Moon's libration (18,19) was constructed on the basis of linear equations whose solutions give a motion close to Cassini's motion.

A classical linear theory is based on the empirical fact that the angle between the axis of the Moon's rotation and the ecliptic axis is small as well as on the smallness of the inclination of the Moon's orbit to the ecliptic.

The development of the classical theories is a quasi-linear theory of the Moon's libration constructed by Habibulin ((20) etc. see also the description of this theory in (21)). This theory embraces a number of weak effects in the linear oscillations of the Moon relative to its center of mass. Here we shall be interested in the other aspect of the problem: the possibility of obtaining from the equations of motion Cassini's laws themselves (and their possible generalizations) as laws of nature.

Cassini's laws are of an empirical nature and do not satisfy the exact equations of motion and that is why they were not derived from the exact non-linear equations of motion. The fact of the existence of the "motion noted by Cassini" points to the stability (in some sense) of this motion. One can expect that there exist approximate, but nearly exact, non-linear equations of motion containing in themselves Cassini's steady motions as stationary ones. Construction of such equations and their analysis is given in (1,2). Investigations on this theme have been also carried out independently by M. L. Lidov and A. I. Neustadt. Earlier the problem was partly investigated by Colombo (30) and Peale (32).

Owing to the success in the investigation of the motion of bodies in the gravitational field achieved recently (22-29), it is possible to construct non-linear equations having the above properties. These equations are equations of the first approximation in the sense of asymptotic methods of non-linear mechanics (the averaged method). While constructing these equations the essential factors determining the Moon's rotational motion are taken into consideration, i.e. the rotation of the plane of the Moon's orbit in space (the secular motion of the Moon's orbit node resulting from perturbations by the Sun); the influence of the Earth's gravitational forces on the Moon; and the resonance of the phenomenon (commensurability of the periods of axial rotation of the Moon and its orbital period). Application of asymptotic methods is justified by the smallness of gravitational torques due to proximity to the sphere of the ellipsoid of the Moon's inertia. Similar facts are taken into account in the case of Mercury.

Stable relative body equilibrium in a circular undisturbed (immovable in space) orbit is a close non-linear model of Cassini's law. The existence and stability of such motion were stipulated by the effects of gravitational torques on the Moon from the Earth and were investigated approximately, based on linear equations, by Lagrange in 1780 (17,18). The existence of the relative equilibrium and its stability in sense was studied rigidly by Lyapunov only in 1956 (26). The result of this investigation can be formulated as follows (22,24): such motion of the body (relative equilibrium in the circular orbit) exists where the main central axes of inertia of the body coincide consistently with the directions of the orbital radius vector, the normal to the orbit plane and the tangent to the orbit. To have stability relative to equilibrium it is enough that the greatest axis of the ellipsoid of inertia is directed along the radius vector, the least axis is directed along the normal to the orbital plane and therefore, the middle axis along the direction of the tangent (Fig. 5).

Relative equilibrium means 1:1 synchronization between the period of axial body rotation and the period of orbital motion.

In relative equilibrium an axis of a body (satellite) is steadily directed to the center of gravity. This fact is used fundamentally by the system of gravitational stabilization of the artificial earth satellite. In this case the satellite is stabilized in such a position that one of its sides turns to the Earth all the time, as with the Moon. This does not sufficiently describe the actual state of affairs. The relative equilibrium satisfies the first of the three Cassini's laws. But two other Cassini's laws are not satisfied.

Cassini's second and third laws are essentially associated with evolution of the Moon's orbit under the influence of external disturbances (in general, solar attraction), i.e. they are associated with the effect of the rotation of the Moon's orbit plane. This effect lies in the fact that the normal to the plane of the Moon's orbit is precessing around the normal to ecliptic with angular velocity (16):

$$K_{\Omega} = -1934^{\circ}08'31'' \text{ for } 100 \text{ years, } 36525 \text{ ephemeris days} \quad .$$

(Thus, the period of the motion of the node ~ 18.6 years.) Cassini's second and third laws mean that in this case the Moon's axis, the normal to the ecliptic and the normal to the Moon's orbit lie in one plane not coinciding with each other (Fig. 4). This is the difference between Cassini's laws and the "law of the relative equation". In the latter the axis of rotation of the satellite (the Moon) coincides with the normal to the orbital plane.

According to Cassini's second and third laws the plane containing the Moon's axis and the normal to the Moon's orbit is always normal to the line of the nodes and therefore rotates with angular velocity K_{Ω} about the normal to the ecliptic. So, the Moon's axis tracks the motion of the normal to the Moon's orbit.

Moreover, an additional 1:1 synchronization takes place between the period of motion (precession) of the Moon's axis in space and the period of precession of the Moon's orbit.

Such motion of the Moon's axis can be treated as its immovability in the reference frame rotating with angular velocity about the normal to the ecliptic.

In 1958-1965 (23-25), Beletskii investigated rapid motion of the satellite under the influence of the gravitational torques taking into account the evolution of the satellite orbit due to external disturbances. The method of investigation is to obtain and analyse some evolutionary equations of motion which are a good approximation to the accurate equations and giving a sufficiently exact qualitative and quantitative picture of the motion. The result of the analysis is as follows: the satellite must move in an undisturbed manner around the constant (in magnitude) vector of angular momentum. This vector itself moves slowly in space. The trajectory of the end of the vector of angular momentum is closed in a rotating reference frame in which one axis is collinear with the line of nodes of the satellite's orbit and another (an axis of rotation) coincides with the normal to the plane of the orbit. The trajectory of the end of the vector of angular momentum is shown in Fig. 6.*

It should be emphasized that the completeness of the trajectory of the vector of angular momentum in the reference frame rotating with the orbit has a tendency to synchronize the rotation of the orbit and the vector of angular momentum. In (23-25) this fact is found with regard not only to gravitational but to magnetic and aerodynamic effects.

---

*In (23-25) the case of a dynamically symmetrical body was investigated, but the result can be slightly generalized for the case of three-axis body (26,27). This paper considers the rapid rotation of a three-axis satellite in the gravitational field. Evolution of the orbit was not taken into account. In some later American publications (30,31,37,38), the effects found in (23-26) are rediscovered.

A special case of the motion described is an axial body rotation about the main central axis of inertia which coincides with the direction $\vec{K}$ in Fig. 6 (or with the direction at the other stationary point). As is clear from the character of the trajectory of the vector of angular momentum in the neighborhood of direction $\vec{K}$ (Fig. 6) this particular motion is stable in the sense of the position of the axis $\vec{K}$. Also, this particular motion "satisfies in the generalized sense, Cassini's second and third laws" as Colombo acknowledged in 1966 (30). In this paper he also stated and discussed an idea concerning the application of Cassini's generalized laws to the motion of other celestial bodies (Mercury, Japet) besides the Moon.

But with this method of investigation the motion given in (23-25) cannot satisfy Cassini's first law. Evolutionary equations were built from the fact of existence of a small parametric relation of the orbital angular velocity to the angular velocity of body rotation. The latter is essentially greater than the former and cannot be equal to it as Cassini's first law requires.

But the small parameter necessary for the construction of the evolutionary equations may be introduced in some other way: we use the fact that the ellipsoid of inertia of the Moon is close to a sphere, i.e. the Moon's moments of inertia are nearly equal. For this case Chernousko (26) constructed and investigated the evolutionary equations (without taking into account orbital evolution). These allow us to consider any angular velocities (not only resonance velocities) of satellite rotation including the velocities close to the vertical. Adding to these equations those developed in (23-25) we find an amount of motion similar to that mentioned above: the satellite moves in an undisturbed manner (with some stipulations) about the constant (in magnitude) vector of angular momentum. The vector itself describes the trajectory in space shown in Fig. 6. Here the angular velocity of axial satellite rotation is equal accurately to the orbital angular velocity. This motion satisfies formally all Cassini's laws.

But the particular motion just mentioned has a formal character due to the breaking of the condition of non-resonance. Being stable relative to Cassini's second and third laws this motion is unstable relative to the first law: any change in the axial angular velocity is conserved so that the Moon, though slowly, but would have turned its back to the Earth after all.

Cassini's first law establishes the resonance of the Moon's axial and orbital angular velocities are in the 1:1 relation. The construction of the evolutionary equations for such motions is more complicated than for non-resonance motion. Earlier constructions ignore resonance of the phenomena discussed.*

In 1963 Chernousko (27) obtained and investigated a useful equation which described the plane resonance rotation of artificial and natural celestial bodies having

*In (32) S. Peale without writing down equations of motion, formally spreads the analysis of (30) over the "resonance" case. From the equation of motion written down and investigated in (1,2) we can see that trajectories of motion of the type shown in Fig. 6 and trajectories of modulus of the vector of angular momentum considered in (32) do not exist.

their ellipsoid of inertia close to a sphere. Then the resonance space satellite motions were investigated (without taking into account the orbital evolution) in the gravitational field by Torgevskii (28) in 1968-69. The disturbed motion of the satellite, according to these investigations, is not so simple now as in the resonance case, but there are stationary motions when the satellite rotates around the main central axis of inertia (the vector of angular momentum is constant in value and coincides with this axis of inertia). The motion of the vector of angular momentum relative to an immovable (non-evolutionary) orbit is analogous to its motion in the non-resonance cases relative to such an orbit. In this case the motion is stable (for the definite conditions) relative to disturbances of axial rotation. It is this fact which was insufficient in the previous non-resonance constructions. Among the stable stationary motions investigated by Torgevskii there is a case of 1:1 resonance between the orbital and axial frequencies for a body with close moments of inertia, which fits our analysis.

Combining the constructions of papers (23-28) and (33) we are naturally led to "motions by Cassini" which are stable (in the sense of the evolutionary equations) relative to all three Cassini's laws. In this case within the asymptotic limit of the orbital precession to zero we obtain the relative equilibrium and condition of stability of relative equilibrium described in (22,24).

Let us note that in the Moon's orbit the perigee motion is essentially greater than the motion of node. The perigee rotates with the angular velocity $K_{\omega} = 4069^{\circ}02'03''$ in 36525 ephemeris days (the period of motion of the perigee is ~ 8.85 years). According to Trushin (29) one can expect that the motion of perigee does not break the stationary state of motions, at least, in the main terms of the disturbed potential. These ideas are described in (1,2).

Theoretical Analysis Results of the Resonance Rotations

Let us describe some results of investigations carried out in (1,2).

Evolutionary (averaged) equations of motions constructed in (1,2) allow two or four types of stationary motions depending on the value of parameters. Ascending or descending nodes of the satellite's equator coincide with the ascending node of the orbit in each of these motions depending on the value of parameters.

All these motions can be called "Cassini-type motions". There are: direct motions when an angle ρ_0 between the vector of angular momentum and the normal to the plane of the orbit is in the vicinity of a value $\rho_0 \sim 0$ (it is a proper "Cassini-type motion"), retrograde motions (ρ_0 is in the vicinity of a value $\rho_0 \sim \pi$) and inverse motions (ρ_0 is in the vicinity of a value $\rho_0 \sim \pi/2$). In direct motion the axial rotation is in the same direction as the orbital rotation; in retrograde motion, in the opposite direction.

The generalized Cassini's laws: Let the rigid body have a three-axis ellipsoid of inertia which is nearly a sphere, the center of mass of the body have mean angular velocity ω_0 relative to the attracting center in the orbit, differing from the Keplerian one by the constant precession of the normal to the orbital plane about the invariable axis of precession and the constant rotation of the great semi-axis of the orbit in its plane. The body is affected by the gravitational torque of the attracting body.

Then among the possible motions of the body there exist motions which in the first approximation (in the sense of asymptotic methods of non-linear mechanics) satisfy the following three Cassini's generalized laws (1,2):

1. A body rotates uniformly around its main central axis; the angular velocity of this rotation is close to one of the resonance values: ω_0 (the Moon), $3/2\ \omega_0$ (Mercury).*

2. The axis of the body's angular velocity and the normal to the orbital plane are at a constant angle ρ_0. At fixed values of the parameters of the body and the orbit there exist two or four possible values of the angle ρ_0.

3. The axis of angular velocity, the normal to the orbital plane and an axis of precession lie in the same plane.

Let us explain these laws. Let A, B, C be the main central moments of inertia of the body, i the constant inclination of the orbit of the celestial body to the ecliptic; K_Ω and K_ω respectively the constant angular velocities of the motion of the node and the motion of the pericenter of the orbit due to disturbance by other celestial bodies. Let L_0 be the value of the modulus of the vector of angular momentum in the stationary motion. Then the angular velocity of the stationary motion is equal to L_0/B and determined by the formula

*There can exist some other stationary resonance motions.

$$\frac{L_0}{B} = \Omega + K_\omega + K_\Omega \cdot \cos(i \mp \rho_0) \quad ,$$

$$\Omega = \frac{n}{2}\,\omega_0 \quad ; \quad n = 2,3 \quad . \tag{6}$$

This is Cassini's first law (refined): the axial angular velocity is equal to the resonance value. It is equal to the orbital velocity in the case of the Moon or 3/2 of the orbital velocity in the case of Mercury (with a small addition to account for the motions of the pericenter and the orbital node). Cassini's third law is slightly generalized: coinciding nodes of the orbit and the equator can have different or identical names.

The angle ρ_0 included in the statement of Cassini's generalized law is determined in terms of parameters of motions as follows. Let us designate:

$$\alpha = \frac{3}{8}\,\frac{\omega_0{}^2 J}{L_0 K_\Omega}\,(A-C) \quad ;$$

$$\beta = \frac{3}{8}\,\frac{\omega_0{}^2}{L_0 K_\Omega}\,\{\,[3B-(2+J)C] + [B-(2+J)A]\,\} \quad ;$$

$$\chi = \frac{\beta}{\sqrt{\sin^2 i + (\alpha + \cos i)^2}} \quad ;$$

$$\cos\rho^* = \frac{\cos i + \alpha}{\sqrt{\sin^2 i + (\alpha + \cos i)^2}}$$

$$\sin\rho^* = \frac{\sin i}{\sqrt{\sin^2 i + (\alpha + \cos i)^2}} \quad . \tag{7}$$

Here $J = J_n(e)$ is only the function of eccentricity (Fig. 7). $n = 2$ corresponds to 1:1 resonance (the Moon), $n = 3$ corresponds to 3:2 resonance (Mercury). In this case ρ_0 is determined from the equations :

$$\cos\rho^* \mp \sin\rho^* \operatorname{ctg} \rho_0 + \chi \cos\rho_0 = 0 \quad . \qquad (8)$$

This equation was considered in (24) in a study of the non-resonance rotation of the artificial satellite with regard to the evolution of its orbit and a number of disturbing factors (gravitational, aerodynamic and magnetic); later results are given in (30,37,38).

The existence of "four" solutions of Eqs. (8) corresponds to the condition

$$\cos^{2/3}\rho^* + \sin^{2/3}\rho^* < \chi^{2/3} \quad .$$

In particular, this condition is complicated if $|\chi|>2$ at any ρ^*.

The existence of "two" solutions of Eqs. (8) corresponds to the condition

$$\cos^{2/3}\rho^* + \sin^{2/3}\rho^* > \chi^3 \quad .$$

In particular, the latter condition is complicated if $|\chi|<1$ at any ρ^*. Values of $|\chi|<1$ mean predominance of effects of orbit evolution over the effects of gravitational torques and the predominance of the effects of gravitational torques corresponds to large values of $|\chi|$.

An example: the Moon

For the Moon we accept (16): $B = 0,88836978$, $A = 0,88800195$, $C = 0,99781798$ (in 10^{35}kg.m^2), $i = 5°08'43''$. The rotational period of the node of the Moon's orbit is about 18.6 years ($K_\Omega = -1934°08'31''$ for 100 years, 36525 ephemeris days); thus, $|K_\Omega|/\omega_0 = 0.00398324$. The period of the perigee approaches 8.85 years ($K_\omega = 4069°02'03''$ for 100 years).

In the formulae of this section we may consider $L_0 = B\omega_0$. Using this data we calculate $\chi \approx -0.218$, $\cos\rho^* \approx 0.970$ and note that the condition of existence of only two stationary motions is fulfilled with the excess. We find that the necessary stationary motion ($\rho_0 < \pi/2$) provides coincidence of the descending node of the Moon's equator with the ascending node of its orbit. Substituting the given numerical data into the coefficients of Eq. (8) (and taking into account the fact that $J = 1$) we can calculate the root ρ_0 of the equation (close to $\rho_0 \sim 0$). It was found that $\rho_0 = 6°39'42''$. Hence, for the angle of inclination of the Moon's axis to the axis of the ecliptic we have the calculated value $\rho_0 - i = 1°30'59'' \approx 1°31'0$. According to observations (Cassini's second law) this angle is equal to $1°32'1$. Good coincidence of the calculated and observed values underlines the high accuracy of the accepted pattern of motion.

Stability: Equation (8) has two roots satisfying stable positions of the rotation axis of the body, or four roots where three satisfy the stable position of the axis and one satisfies its unstable position. Denote the only possible unstable value ρ_0 in terms of ρ_0^N.

The main results of the analysis carried out for stability of 1:1 resonance motion are as follows. Necessary, close to sufficient, conditions of stability are represented: 1) Rotation takes place around the smallest axis of the ellipsoid of inertia of the body. 2) The orbital radius vector is along the largest axis. 3) The inclination ρ_0 of the axis of rotation to the normal to the orbital plane satisfies the inequality $3+2\sqrt{6}/15 < \cos\rho_0 < 1$. 4) $\rho_0 \neq \rho_0^N$. For the Moon these four conditions are fulfilled.

As follows from the evolutionary equations the motion of the axis in the neighborhood of its stationary position is conditionally periodic with two frequencies $\omega_1 \sim \sqrt{\varepsilon}$ and $\omega_2 \sim \varepsilon$ where ε is an accepted order of smallness of coefficients in the right-hand sides of the evolutionary equations. Thus, the trajectories described by the body's axis on the celestial sphere are not closed in general.

It should be noted that according to the data (3,5,6) Jupiter's satellites Io, Europe, Collisto (JI, JII, JIV) and Saturn's satellite Japet (SVIII), as well as possibly Rhea, Tehys, Dione, and Mars' satellites Phobos and Deimos, rotate like the Moon with 1:1 resonance. That is why the theory constructed can also be used for them. One can expect that due to tidal effects these satellites, like the Moon, have stationary "Cassini-type motions" (in the generalized sense of this paper).

In (1 and 2) the following conditions of stability of 3:2 resonance motion as in Mercury are constructed: 1) $e \neq 0$. 2) Rotation takes place around the smallest axis of the ellipsoid of inertia. 3) The largest axis of the ellipsoid of inertia at each passage of the pericenter is projected on the direction of the large orbital semi-axis. 4) $\rho_0 \neq \pi$; $\rho_0 \neq \rho_0^N$. Let us also note the fact that at the fixed moments of inertia the width of the resonance zone, i.e. "regions of stability", is determined by a value of order of $\omega_0 \sqrt{J_n}$; very small values of J_n may mean that probability for the given resonance is very small. It is found that, at least, according to this point of view Mercury has very little chance of 3:2 resonance compared with the Moon's for 1:1 resonance. This follows from relation:

$$\Delta = \frac{(\Delta_{max})_{merc}}{(\Delta_{max})_{moon}} = \left(\frac{T_{moon}}{T_{merc}}\right) \sqrt{\frac{J_3}{J_2}} = \frac{28}{88} \sqrt{\frac{0.714}{0.993}} \approx 0.25 \tag{11}$$

where Δ_{max} means the width of the resonance region respectively for Mercury and the Moon.

Notes on Resonance in the Rotation of Venus

Theoretical substantiation of the resonance of the rotation of Venus described in the introduction has been discussed in (3).

It was shown that steady resonance of the type mentioned is really possible and at resonance (0:1), interesting for us, the Earth's effects are dominant and the Sun's influence (K_2) is negligibly small on the average.

We can give some reasons which do not speak well for the real existence of this resonance.

1) Resonance (0:1) is not better than many adjacent resonances with the same width of the resonance zone. Therefore it is difficult to explain an assignable

preference for this resonance.

2) The relative width of the resonance zone Δ_v of Venus (comparing with the width of the resonance zone Δ_{moon} of the Moon and Δ_{merc} of Mercury) is negligibly small:

$$\Delta = \frac{(\Delta_{max})_v}{(\Delta_{max})_{moon}} \simeq 0.225\times10^{-3} \quad . \tag{12}$$

3) The analysis carried out in (3) does not satisfy us because of the neglect of the great effect of the Sun. On the average the Sun's influence is really negligible but the amplitude of the periodic (averaged in (3)) disturbances from the Sun may be found to be essentially significant because the Sun's gravitational torques are 10^6 times greater than those of the Earth.

More detailed analysis of this problem (40) reveals the following facts. In order that the periodic solar disturbances do not withdraw motions from the resonance zone the ratio of the moments of inertia of Venus must be situated within the limits:

$$\frac{A-C}{B} \lesssim 2.5\times10^{-5} \quad . \tag{13}$$

The values of the moments of inertia of Venus are unknown, but if we remember that for the Moon the analogous ratio has an order of $\sim 5\times10^{-4}$ we can conclude that the requirement stated falls within the boundary of reasonable limits. Therefore it is considered that the investigated resonance in Venus' rotation can be just on the verge of possibility.

Let us also note that according to the analysis in (3, 40) Venus' capture in resonance (0:1) cannot be explained by tidal phenomena; it is necessary to build other models of dissipative phenomena to explain it. This is the fourth argument which does not speak well for the regular (assignable) existence of resonance (0:1).

Thus, resonance in Venus' rotation is not indisputable as are the resonancesof the Moon and Mercury. If, nevertheless, it exists, then we have the right to call the Venus' resonance rotation a natural phenomenon.

Tendency to Synchronization of the Rotational Motion in Compound Gravitational Fields. Luni-Solar Precession and Nutation of the Earth's Axis

So far we have considered the rotational motion of a body as the dominating influence of the gravitational field which is the only one attracting the center. But in Nature some situations can appear which require us to take into account a great number of attracting centers compared with the value of their influence. The classical example of such a situation is a disturbed motion of the Earth's axis. The attractions of the Sun and the Moon make comparable contributions to this motion.

The disturbed motion of Earth-luni-solar precession and nutation is not resonance in the sense that the Moon and Mercury motions are resonance. It should be noted here that in the Earth's motion there exists a tendency to synchronization. Let us clear up this requirement.

Let C and A be the longitudinal and transverse moments of inertia of the Earth; m_{moon} and m_{ϱ} are masses of the Moon and the Earth respectively; e is the eccentricity of the Moon's orbit; e_e is the eccentricity of the Earth's orbit; θ_0 is the angle between the Earth's axis of rotation and the Earth's vector of angular momentum (to high accuracy $\theta_0 \approx 0$); L_0 is the modulus of the Earth's vector of angular momentum; $\dot{\Omega}$ is the angular velocity of the precession of the node of the Moon's orbit. Let ω_e be the average angular velocity of the Earth's orbiting and ω_{moon} the average velocity of the Moon's orbiting. Let us introduce the constant parameters

$$K_C = \frac{3}{8}\frac{A-C}{L_0}(3\cos^2\theta_0-1)\frac{m_\Lambda}{m_\Lambda+m}\frac{\omega_\Lambda}{(1-e^2)^{3/2}} \quad ;$$

$$K_\odot = \frac{3}{8}\frac{A-C}{L_0}(3\cos^2\theta_0-1)\frac{\omega_3^2}{\omega_\Lambda(1-e_3^2)^{3/2}} \quad ;$$

$$K_\Omega = \frac{\dot{\Omega}}{\omega_\Lambda} \quad , \tag{14}$$

and construct the function

$$\Phi = -L_0K_C\cos^2\rho - L_0K_\odot\cos^2\Delta + L_0K_\Omega\cos\Delta \quad , \tag{15}$$

where ρ is the angle between the vector of angular momentum and the normal to the plane of the Moon's orbit and Δ is the angle between the same vector and the normal to the plane of the ecliptic (the Moon's orbit is inclined to the ecliptic at a constant angle i). In virtue of the evolutionary equations of motion the angles ρ and Δ are variable in time, but from these equations there follows the existence of the integral

$$\Phi = \Phi_0 = \text{const.} \tag{16}$$

and from (16) there follows immediately the fundamental fact:

"All possible trajectories of the vector of angular

momentum are closed in the reference frame rotating with the plane of the Moon's orbit."

In the classical luni-solar precession and nutation of the Earth axis this fact (and integral (16)) is evidently not found, but it reflects the tendency to synchronization of the precession of the Earth's axis and the precession of the Moon's orbit. Actually, many closed trajectories shrink to special points satisfying the stationary motions. These stationary motions can be interpreted as the next generalization of Cassini's laws in the case of influence of many centers of attractions (here of two). In the process of evolution under the influence of dissipative factors the motion had to be shrunk to a stationary one where the Earth's axis is precessing with the same velocity as the plane of the Moon's orbit is precessing. The real situation, of course, is far from the stationary one described above (the period of precession of the Moon's orbit is 18.6 years and of the Earth's axis 26000 years), but we are interested in the principle of the matter.

The actual nutation of the Earth's axis with period 18.6 years is the reflection of the tendency to synchronization.

Model of Tidal Phenomena and Capture into Resonance Rotation

It has been shown that the gravitational torques acting on a celestial body can guarantee its stable resonance motion as described by Cassini's generalized laws. But the initial data of motion can be situated outside a resonance zone: evidently, a probability of the initial into the resonance zone is small. Therefore a mechanism must exist which gives sufficiently high possibility of motion capture (in the process of its evolution) into the resonance zone if the initial data are outside this zone. Tidal phenomena are, evidently, such a mechanism. The theory of tidal phenomena is complex and depends on many factors; here we limit ourselves to consideration of a few models.

In (40) a model of a tidal moment is presented. Let us consider a case in space. We designate $\vec{e}_\omega$ as the unit vector of the planet's angular velocity, $\vec{e}_r$ as the unit vector in the direction from the planet to the center of attraction. Then the vector of tidal momentum $\vec{M}$ is given by

$$\vec{M} = \frac{k \sin 2\delta}{r^6} \left[\vec{e}_\omega \times \vec{e}_r\right] \times \vec{e}_r \quad . \tag{17}$$

Here k = const, r is the direction from the center of attraction to the center of mass of the planet; δ is the angle of shift of the tidal delay relative to the direction $\vec{e}_r$.

The hypothetical dependence of $\delta(\omega)$ is shown in Fig. 8. This dependence determines two usually used (3,35) analytical models for $\delta(\omega)$:

$$(1) \quad \sin 2\delta \sim 2\delta \sim k_1\omega \quad ; \quad k_1 = \text{const} \tag{18}$$

$$(2) \quad \sin 2\delta = \sin 2\delta_0 \quad ; \quad \delta_0 = \text{const} \tag{19}$$

Model Eq. (18) involves a wider range, but it is very approximate in the neighborhood of small values of the parameters ω and δ.

The analysis of the evolutionary equations of the rotation of the planet under the influence of a tidal moment (Eq. 17) shows that rotation tends to be axial. Let $\Omega = L/A\omega_0$ be the angular velocity of the rotation of the planet referred to the angular velocity ω_0 of its orbital motion; ρ is the angle between the vector of angular momentum $\vec{L}$ and the normal to the plane of the orbit. Integral curves $\rho(\Omega)$ are shown schematically in Fig. 9. We can see that the motion tends to the only stationary point

$$\rho = 0 \quad ; \quad \Omega = \Omega^* = \frac{1 + \frac{15}{2} e^2 + \frac{45}{8} e^4 + \frac{5}{16} e^6}{(1-e^2)^{3/2}(1+3e^2 + \frac{3}{8} e^4)} . \tag{20}$$

Thus the tidal moment tends to cause motion by direct rotation around the normal to the plane's orbit with an angular velocity corresponding to Eq. (20). One can see that the tidal moment always draws out the motion, from a retrograde ($\rho \sim \pi$) rotation. This is the fact that is interesting for the theory of the rotation of Venus. Its retrograde rotation cannot be explained by capture under the influence of tidal moments.

The fact of dependence of the limiting angular velocity on the orbital eccentricity is also interesting for us. The function $\Omega^*(e)$ is illustrated in Fig. 10. Here at $e = 0$ $\Omega^* = 1$, i.e. in the circular orbit the rotation is limited to 1:1 resonance (relative equilibrium). Colombo and Shapiro (35) investigated, from this point of view, Mercury's motion and found that its capture into resonance motion with 3:2 resonance can be established by eccentricity $e \simeq 0.2$. In our model $\Omega^* = 3/2$

if $e \approx 0.28$; other models may decrease the required value for e. For example, assuming Eq. (19) the limiting value $\Omega^* = 3/2$ is obtained for $e = 0.25$. In general the spectrum of values of eccentricities at which the given value Ω^* is obtained under the influence of tidal forces (Table 1) corresponds to the spectrum of the resonance values Ω^*.

Table 1. Values of eccentricity e for various values of Ω^*.

Ω^*	1	1.5	2.0	2.5	3.0	3.5
Model Eq. (18)	0	0.284	0392	0.465	0.519	0.561
Model Eq. (19)	0	0.246	0321	0.359	0.385	0.404

In Table 1 the second line is calculated according to formulas (18) and (20) and the third line is calculated according to the formula for model (19):

$$\Omega^* = 1 + \frac{19}{2} e^2 - \frac{433}{8} e^4 - \frac{43727}{80} e^6 + \dots \qquad (21)$$

Formula (21) is approximate, and (20) accurate.

If besides the tidal phenomena we take into account the gravitational torque studied earlier formulae (20), (21) will have another meaning. Orbits having eccentricities given in Table 1 provide for the maximum possibility of capture ($P \sim 1$) in the corresponding resonance. If the eccentricity is close to the value given in Table 1 the probability of capture is however, less than 1, but it is still great. This is obtained from Eqs. (20), (21) and the spectrum of eccentricities given in Table 1.

In Eq. (18) the calculation of the probability (3,40) of capture for the Moon ($e \simeq 0.05$) in 1:1 resonance and for Mercury ($e \simeq 0.2$) in 3:2 resonance gives for the Moon $P \simeq 0.6$, for Mercury $P \simeq 0.25\times10^{-1}$. Thus the capture probability for the Moon is great. The capture probability for Mercury is not very great but it is finite. The capture probability for Venus is calculated as $P \simeq 0.5\times10^{-4}$.

In other words the probability of the capture of Venus in the resonance rotation by tidal forces is negligible. It should be also noted that calculation of capture probability was carried out for the plane model of motion. For Venus such an estimate is problematic because of the instability of the plane rotation in the presence of tidal forces compared with that due to space disturbances.

The Moon and Mercury have not this defect in virtue of the direct rotation; other models of the tidal effects can increase the estimate of capture probability (for Mercury it is of order of 3).

Table 2 gives the resonance motion of the Moon, Mercury and Venus. In that Table T_0 is the period of orbiting (v = Venus, e = Earth); T_ω is the period of rotation; τ is the period of the system Venus-Earth; Δ_{rel} is the relative width of the resonance zone; e_{cap} is the value of eccentricity of the orbit which provides for capture in the resonance considered with probability $P = 1$; e is the actual eccentricity; P is the probability of capture in the corresponding resonance by tidal forces (according to the model of the tidal effects considered above); ρ is the inclination of the axis of rotation of a celestial body to the normal to its orbit.

Table 2. Resonance Motion of the Moon, Mercury and Venus

1	Moon 2	Mercury 3	Venus 4
T_0 (days)	27.32	87.97	$T_0^V = 224.7$ $T_0^e = 365.25$ $\tau = 583.92$
T_ω (days)	27.32	59±3	243±0.1
direction of motion	direct	direct	retrograde
resonance	$\left(\frac{T_\omega}{T_0} - 1\right) = 0$	$\left(\frac{3T_\omega}{2T_0} - 1\right)$ $\simeq \pm 0.017$	$\left(\frac{4}{T_0^V} - \frac{5}{T_0^e} - \frac{1}{T_\omega}\right) T_\omega$ $= \pm 0.001$
Δ_{rel} (relative)	1	0.25	0.225×10^{-3}
e_{cap} (capture)	0	0.25-0.28	-
e	0.055	0.206	0.007
P	0.6	0.25×10^{-1}	0.5×10^{-4}
ρ (grade)	6°40	<7°	1°2

The Satellite's Magneto-Gravitational Stabilization

An example of the use of resonance effect is the idea of magneto-gravitation stabilization for artificial satellites (41). For satellites with long useful life it is desirable that they be oriented in space by passive methods using external torques. For example, the interaction of the satellite's own magnetic field and the Earth's magnetic field can be used to create a system of orientation by magnetic field.

If we approximate the geomagnetic field by a dipole whose axis coincides with the axis of the Earth's rotation, the magnetic field vector H changing at each point of the orbit with inclination 1 will trace out conic surfaces in space which are approximately circular cones with vertical angle 2ν where $\nu \leq \pi/2$.

Due to the irregular motion of the vector H it is impossible to obtain an ideal passive orientation by the magnetic field. Hence we can only strive for such orientation on the average. It is known that, using, for example, gravitational (magnetic) and other torques for the passive stabilization of torques of a certain type, the magnetic (gravitational) torques appear as disturbing factors which cause the satellite's departure from normal motion.

Paper (41) shows that in some cases gravitational torques can favor the satellite's orientation by the geomagnetic field in coordination with magnetic torques. This is based on the following considerations: among the stable motions of a satellite in the gravitational field there is one which is close (in the sense of asymptotical methods of the theory of oscillations) to resonance precession with a 2:1 relation between the frequency of precession and the frequency of revolution. An analogous motion exists if a magnetized satellite is affected by simultaneous magnetic and gravitational torques. In this case the interaction of the two fields creates stationary trajectories of the vector of angular momentum: this vector preserves a steady immobility. Thus, the 2:1 type of resonance precession can take place around an axis immovable in space. In this motion there are two free parameters: the vertical angle of the cone of precession and the relation between the moduli of the magnetic and gravitational torques. On the other hand the magnetic field vector in the dipole approximation describes a conic surface in space with the same relation of frequencies (two turns by the cone for one satellite revolution). Choosing the free parameters mentioned above in the required way, the cone of precession can be made to coincide with the cone of the magnetic field; this gives the alignment in the average, of the satellite axis by the magnetic field vector. Thus, the satellite can be stabilized on average by the Earth's magnetic field by means principally of the joint effects of the gravitational and magnetic torques.

Evidently, such effects can also play a certain role in the rotation of some natural satellites of the planets if these planets and their satellites have their own magnetic fields. As has already been mentioned a number of natural satellites (the Moon, several of Jupiter's satellites and Saturn's satellites, and evidently both Mars satellites) rotate with 1:1 resonance. Mercury rotates with 3:2 resonance. Motions of these bodies are subjected to Cassini's generalized laws which are the result purely of the influence of the gravitational torques on the body. The effect of the magnetic torques can favor the appearance of 2:1 resonance in the rotational motion of the natural celestial bodies.

To determine the satellite's position let us introduce (24) the variables L, ρ, Σ, ψ, ϕ, θ. Here L is the modulus of the satellite's vector of angular momentum $\vec{L}$ relative to the center of mass, ρ and Σ are the angles giving the position of the satellite's principal axes of inertia xyz relative to the axes LL_1L_2 bounded with the vector $\vec{L}$. Let u be the argument of the latitude, μ is the gravitational constant of the Earth; A,B,C are the satellite's moments of inertia relative to the axes x, y and z, μ_e is a constant of the Earth's magnetism. Let the satellite's magnetic torque be constant and J be a value of its projection on the axis z. We shall consider that the orbit is circular and does not depend on the rotational motion of the satellite. The satellite's moments of inertia are similar and A=B, the force functions U^P and U^m of the gravitational and magnetic fields have the same order and in addition $|U| << L^2/A$.

Then in the system of evolutionary equations the variables ψ and u will vary rapidly and the remainder will vary slowly. If a satellite is oriented by the geomagnetic field its angular velocity will be close to the double velocity of revolution. Therefore to analyze the motions close to such regimes of stabilization we substitute the variables $\psi = 2u + \kappa$ (κ is a new slow variable) and average the right hand terms of the equations (that is equivalent to the average of the force function U) with respect to the fast variable u.

We obtain the system of the averaged evolutionary equations (41).

This system has a particular solution in form

$$\rho = \rho_0 , \quad \Sigma = \Sigma_0 , \quad \kappa = \kappa_0 , \quad \theta = \theta_0 , \quad L = L_0 \tag{20}$$

Solutions in which an axis of symmetry of the satellite moves uniformly with angular velocity $2\omega_0$ on the surface of a circular cone with vertical angle 2ν coinciding with the geomagnetic field vector at the value of argument of the width to $k \cdot \pi/2$ ($k = 0,1,2$) are of special interest, the axis z is tracking on the average the vector $\vec{H}$, and the vector $\vec{L}_0$ is collinear with the cone's axis of symmetry. These solutions satisfying the conditions

$$\rho_0 = \nu - I, \; \Sigma_0 = \kappa_0 = 3\pi/2, \; \theta_0 = \nu ,$$

$$L = L_0 = 2A\omega_0 + 0\left(\frac{|U|A}{L_0^2}\right) \tag{22}$$

exist if the non-dimensional parameter $\zeta = J\mu_e/3\mu(A-C)$ take the values shown in Fig. 11.

As is seen from Fig. 11 the solutions exist for the majority of inclinations (excluding orbits close to equatorial ones where the required value ζ becomes very great. (We must remember that according to the statement of the problem U^P and U^m are compared.) It is interesting that these solutions exist only in the presence of gravitational torques (if $A \to C$, then $\zeta \to \infty$). Thus it is found that the gravitational torques help to stabilize a satellite by the geomagnetic field.

Investigating the stability of the stationary motions found, one can show that the necessary and sufficient conditions of stability are fulfilled if $A<C$, $J>0$. The first of these inequalities means that the geomagnetic field vector is steadily tracked only by the smallest axis of the satellite's ellipsoid of inertia.

The width of the resonance zone is determined by the value $\{3(C-A)\beta/2A\}^{1/2}$ in the first approximation (in the case of the polar orbit this estimation coincides with the accurate one). The diagram of dependence of the non-dimensional parameter β on the orbital inclination is shown in Fig. 12. We can see that for the majority of inclinations the resonance zone is not narrower than the one found strictly for the polar orbit and the resonance effect is great.

Gravicraft

The resonance state between the rotational and translational movement of a body can lead to qualitatively new effects. In 1967 V. V. Beletskii and M. E. Guivertz offered an idea of "gravicraft" (42,43), a new and somewhat unexpected method of manoeuvre in cosmic space. Strange regularities of resonance phenomena are brilliantly expressed in this idea. It is based on the following points:
1) A force of attraction acting on a body of finite sizes differs from the force acting on a point of the same mass concentrated at the center of mass of the body. A force of attraction acting on a stretched body depends on its size, form and orientation.
2) By varying the sizes and the form of a body or its orientation we can change the value of the force of attraction acting upon it.
3) One can control this change in the body's size and form (or orientation) in such a way that variations of the force of attraction can lead to a trajectory of the body's motion essentially different from the initial one.

The control must be principally resonance for this case.

The frequency of variation in the gravicraft's size (or in its orientation) must be commensurable with that of its orbital motion. It seems (42) that we can change the orbit from circular to parabolic.

The additional accelerations appearing in this way are of the order of the square of the ratio of the gravicraft's size to the distance from it to the center of attraction. In other words, these accelerations are very small for apparatus of the usual sizes. To make the effect appreciable the gravicraft must have a size equal to hundreds of kilometers. Thus the idea of "gravicraft" has only theoretical interest, nevertheless, now this idea is widely discussed in literature.

In conclusion, after A. M. Molchanov (4) and I. I. Blehman (44), I want to emphasize the generality of phenomena of synchronization (resonance) of the different material objects in nature and technology and to conclude this review by quotation taken from my book (43). "Everything discussed speaks in favor of the specific and assignable role of resonance in the World's motion."

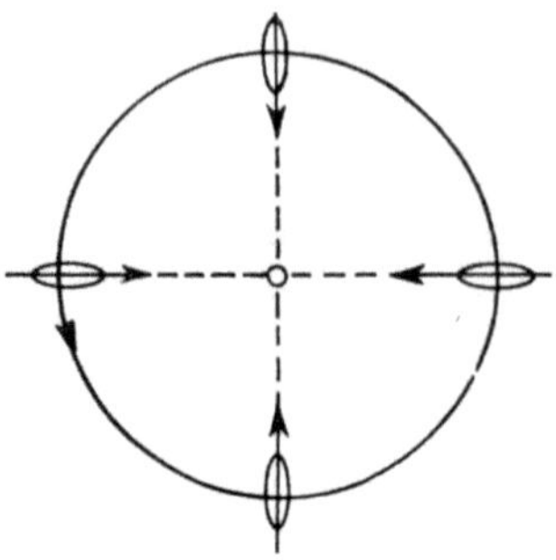

Figure 1.

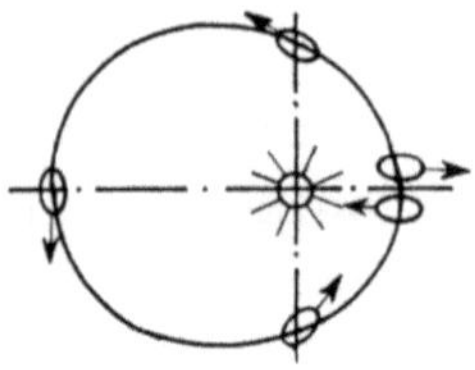

Figure 2.

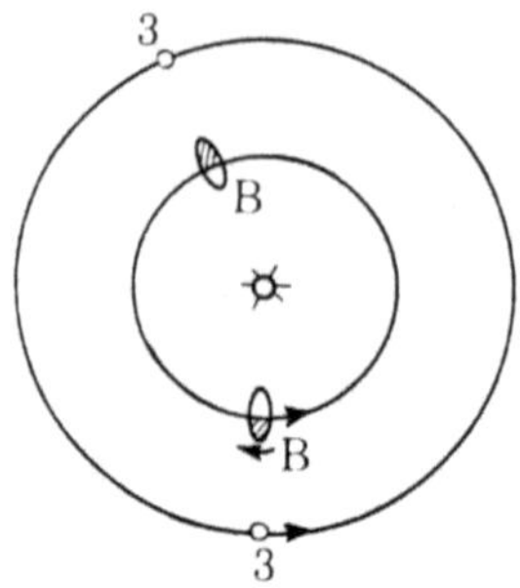

Figure 3.

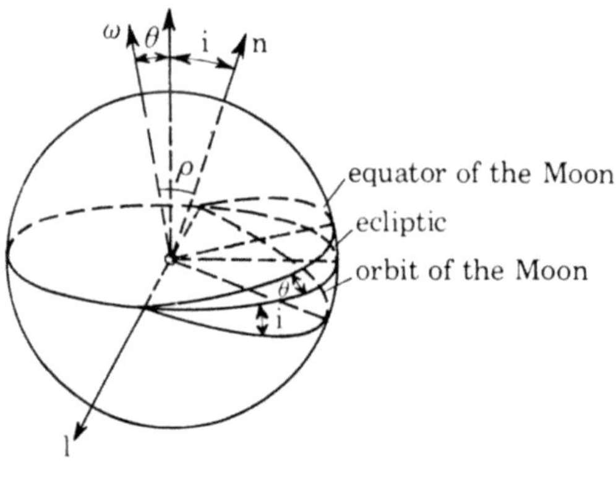

Figure 4.

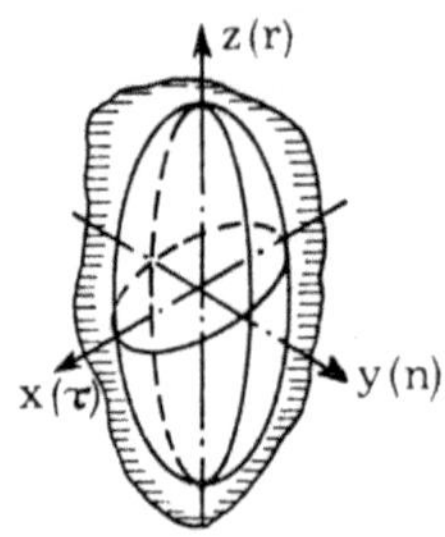

Figure 5.

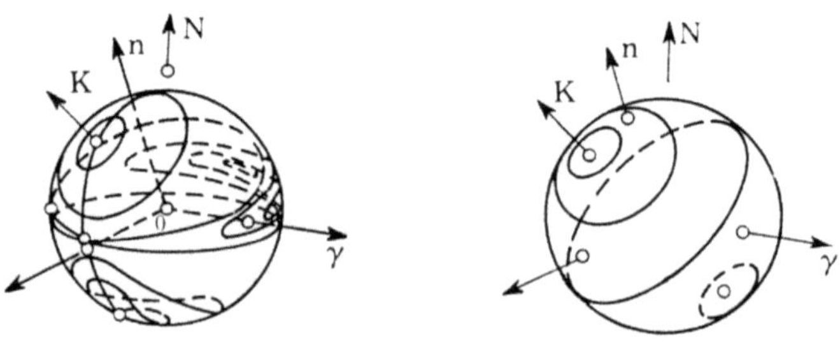

Figure 6.

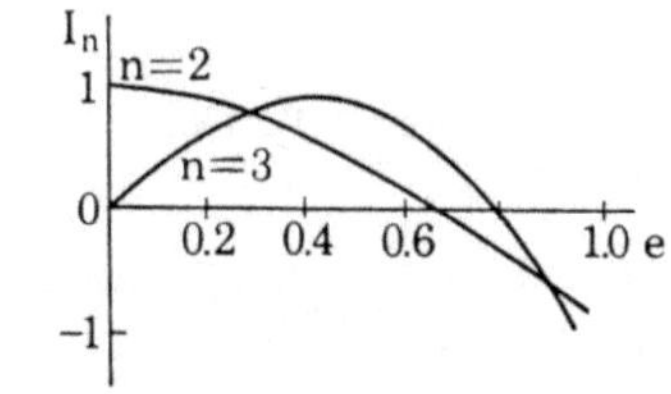

Figure 7.

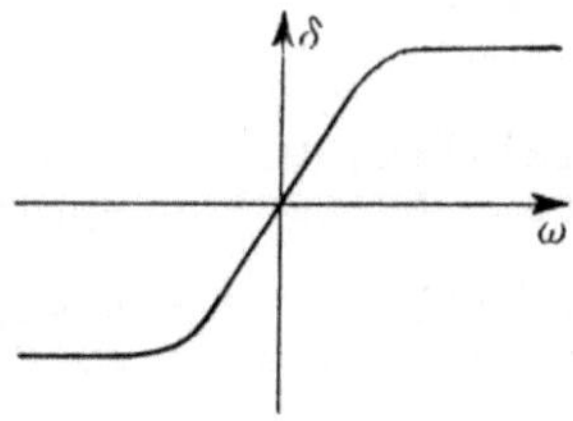

Figure 8.

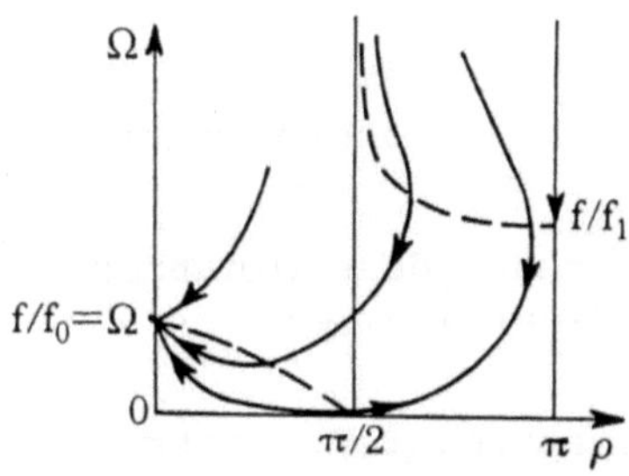

Figure 9.

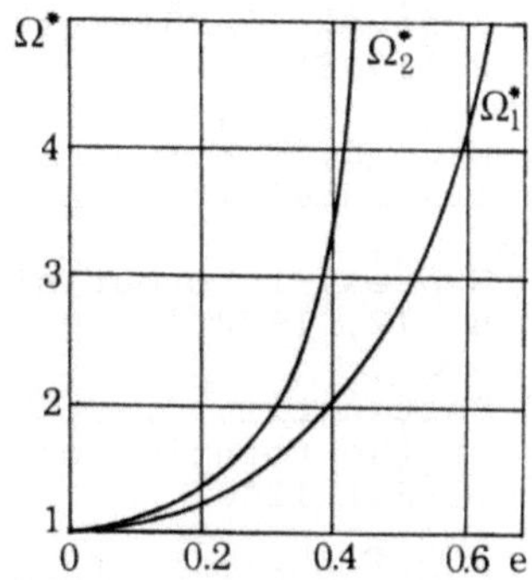

Figure 10.

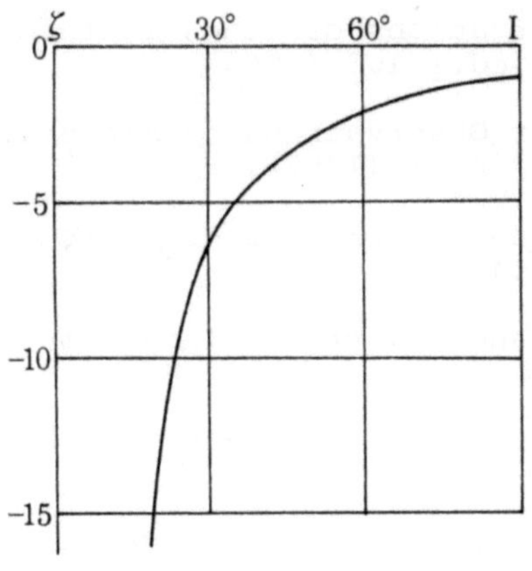

Figure 11.

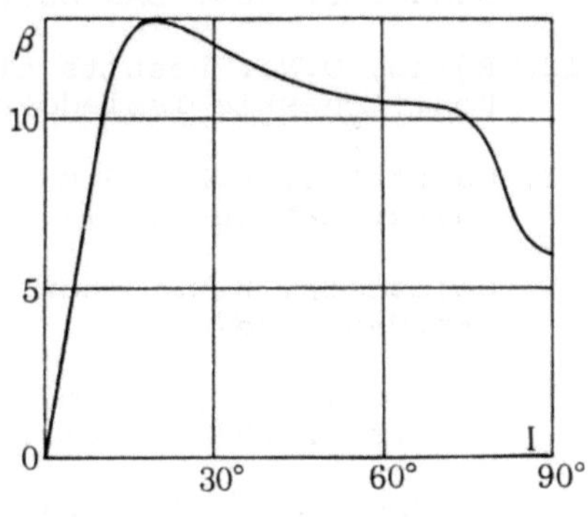

Figure 12.

References

1 Beletskii, V.V.: The Cassini's Laws. Inst. Prik. Matem. AN SSSR preprint, n. 79 Moscow (1971).

2 Beletskii, V.V.: Resonance Rotation of Celestial Bodies and Cassini's Laws. Celes. Mech. 6,n. 3 (1972).

3 Goldreich; Peale, S.: The Dynamics of Planetary Rotations. Annual Review of Astronomy and Astrophysics. 6, Palo Alto, California, 1968.

4 Moltchanov, A.M.: The Resonant Structure of the Solar System. ICARUS-Intern. J. of the Solar System 8, n. 2 (1968).

5 McCord, T.B.; Johnson, T.V.; Elias, J.H.: Saturn and its Satellites. Astrophys. J. 165 (1971) 413.

6 Express-informatisia. "Astronavtika i raketodinamika." n. 12, 1972.

7 Pettengill, G.H.; Dyce, R.B.: A radar determination of the rotation of the planet Mercury. Nature n. 206 (1965) 1240.

8 Dyce, R.B.; Pettengill, G.H.; Shapiro, I.I.: Astron. J. 72 (1967) 351.

9 Colombo, G.: Rotational Period of the Planet Mercury. Nature n. 208 (1965) 575.

10 Beletskii, V.V.: Letter to the editor of Fisica v Schkole. n. 6 (1948).

11 Kotelnikov, V.A.: Radar Observations of Venus in the USSR for 1964. DAN SSSR (1st ed.) 163 (1965).

12 Rjiga, O.N.: Results of Radar Observations of Planets. Kosmitcheskie Issledovania 7, n. 1 (1969).

13 Carpenter, R.L.: Study of Venus by CW radar-1964 results. Astron. J. 71 (1966) 142.

14 Goldstein, R.M.: Moon and Planet. North-Holland, Amsterdam, 1967.

15 Goldreich, P.; Peale, S.: Astron. J. 71 (1966) 425; Nature n. 209 (1966) 1117.

16 Handbook of Celestial Mechanics and Astrodynamics. Edited by G. N. Dubochin. Nauka, Moscow, 1971.

17 Tisserand, F.: Traité de mecanique celeste, tome II, Paris, 1891.

18 Lagrange, J.L.; Oeuvres, Tome V.; Gauthier-Villars, Paris, 1870.

19 Laplace, P.S.: Mecanique celeste, Paris, 1793.

20 Habibulin, Ch.T.: Trudi Kazanskoi gor. astr. observat., n. 34, Kazan, 1966.

21 Kulikov, K.A.; Burevitch, V.B.: Principles of Lunar Astronomy. Nauka, Moscow, 1972.

22 Beletskii, V.V.: The Librations of Satellites. In Iskustvenie sputniki zemli. AN SSSR (3rd ed.) (1959) 13-31.

23 Beletskii, V.V.: The motion of a satellite with respect to its center of mass. In Iskustvenie sputniki zemli. AN SSSR (6th ed.) (1961) 11-32.

24 Beletskii, V.V.: The motion of a satellite with respect to its center of mass. Nauka, Moscow, 1965.

25 Beletskii, V.V.: The motion of a satellite with respect to its center of mass. In Iskustvenie sputniki zemli. AN SSSR (1958) 25-43.

26 Tchernousko, F.L.: The motion of a satellite with respect to its center of mass by the action of gravitational moments. PMM (3rd ed.) 27 (1963) 474-483.

27 Tchernousko, F.L.: Resonance effects in the motion of a satellite with respect to the center of mass. Jornal Vitchisl. Matem. i Matem. Fis., 3, n. 3 (1963) 528-538.

28 Torgevskii, A.P.: The motion of a satellite with respect to the center of mass and resonance effects.

29 Trushin, S.I.: The effect of the motion of the perigee in the problem of the rotation of a satellite. In Kosmitch. issled. 8, n. 4 (1970) 628-629.

30 Colombo, G.: Cassini's Second and Third Laws. The Astron. J. n. 1344 (1966) 891.

31 Holland, R.L.; Sperling, H.J.: A first-order theory for the rotational motion of a triaxial rigid body orbiting and oblate primary. Astron. J. 74, n. 3 (1969).

32 Peale, S.: Generalized Cassini's Laws. The Astron. J. n. 1368 (1969) 483.

33 Beletskii, V.V.: The motion of a dynamically symmetric satellite. In Kosm. Issled. 1 (1963) 339-385.

34 Lutz, F.H.; Abbiz, M.Z.: Rotation locks for near-symmetric satellites. Celes. Mech. n. 1 (1969).

35 Shapiro, I.I.: Radar Astronomy, General Relativity and Celestial Mechanics. In "Modern Questions of Celestial Mechanics," Bressanone, 1967, Coordinator: Prof. G. Colombo. Edizioni Gremonese, Roma, 1968.

36 Lidov, M.L.; Neichtadt, A.I.: The method of the canonic transforms and the Cassini's laws in the problem of the rotation of celestial bodies. Institut prikladnoi matematiki AN SSSR, preprint n. 9, Moscow, 1973.

37 Johnson, D.B.: Precession Rate Matching a Space Station in Orbit about an Oblate Planet. Journal of Spacecraft and Rockets 10, n. 7 (1973).

38 Hara, M.: Effect of Magnetic and Gravitational Torques on Spinning Satellite Attitude. AIAA J. II, n. 12 (1973).

39 Beletskii, V.V.; Truchin, S.I.: The resonances in the rotation of celestial bodies and the generalized Cassini's laws. In Mekanika tverdovo tela. 6th ed.: "Naukova dumka," Kiev, 1974.

40 Beletskii, V.V.; Truchin, S.I.: Stability of the generalized Cassini's laws. In Mekanika tverdovo tela. 6th ed.: "Naukova dumka," Kiev, 1974.

41 Beletskii, V.V.; Heitov, A.A.: Magneto-gravitational stabilization of a satellite. AN SSSR, n. 4, "Mekanika tverdovo tela," (1973).

42 Beletskii, V.V.; Guivertz, M.E.: The motion of an oscillating rod subjected to a gravitational field. In Kosmitcheskie Issledovania 5, n. 6 (1967).

43 Beletskii, V.V.: The motion of celestial bodies. Nauka, Moscow, 1971.

44 Blehman, I.I.: Synchronization of dynamic systems. Nauka, Moscow, 1971.

The Rotational Motion of a Satellite and Atmospheric Density Determination

V.V. BELETSKII, V.M. GRIGOREVSKY, S.Ya. KOLESNIK
Moscow

Abstract

The effect on a rotating satellite of aerodynamic moments resulting from collision with air molecules is considered. Both the diffuse reflection and mirror reflection of molecules from the satellite surface are taken into account. The fact that the atmosphere density may be dissimilar in different altitudes even within the satellite's length, is also taken into consideration.

Under these conditions a general expression is derived for the moment of aerodynamic forces. It is shown to be the sum of three moments associated with the orbital motion of the satellite, its rotation and the existence of the atmospheric density gradient. Let us call them the moments of aerodynamic pressure, aerodynamic friction and "aerogradient".

The analysis of these for a satellite having geometric and dynamic axes of symmetry and a plane of symmetry perpendicular to the axis of symmetry has shown that the moment of aerodynamic pressure can produce an autorotation - a peculiar twisting of the satellite that causes an acceleration of the satellite's spin. The spin can be also accelerated by the moment of aerogradient. The integration of the equation of satellite motion allowing for the moment of aerodynamic friction has enabled us to derive a formula determining the atmosphere density from the satellite drag.

Comparative estimations of the effect of each of the above moments have been carried out, and values of atmosphere density have been computed taking account of aerodynamic mass center (and the influence of other moments) are extensively considered in (1).

Introduction

An earth's satellite is exposed to a number of forces which perturb its motion around the mass center. One can distinguish the following principal types of moments of these forces:

1) Aerodynamic moments; 2) magnetic moments; 3) gravitational moment; 4) moment of light pressure. It should be noted at once that the moment of light pressure forces is of significant importance for balloon satellites whose mass-cross-section area ratio is rather small; for massive satellites however, the moment of light pressure forces is negligible compared with the first three moments.

In the present work the dynamics of a rapidly spinning satellite is studied by the method of averaging. Each of the above moments (except the gravitational one) after time averaging can be resolved into two components of which one is parallel and the other perpendicular to the vector of the satellite's angular moment.

The parallel component makes the value of angular moment vary; consequently the satellite's spin velocity is also changed. Let us call this an accelerating component. The perpendicular component causes variation in the direction of angular moment without varying its value, i.e. simply varies the orientation of the satellite's rotational axis. Let us call this a conservative component (2). The Earth's gravitational field is conservative, and the gravitational moment has no component parallel to the vector of the angular moment. The influence of the gravitational moment on the satellite's motion around the mass center (and the influence of other moments are extensively considered in (1)).

This paper is concerned with the determination and comparative estimation of the accelerating components of aerodynamic and magnetic moments under assumptions more general than in (1). These include assumptions on satellite orientation, the character of molecular impact with its surface, and the reciprocal disposition of axes of geometric and dynamic symmetry depending on Sputnik construction peculiarities. The problem is considered on the basis of work (1,2,4,13) and lectures on the rotation of a satellite delivered by V. V. Beletskii at the Moscow State University throughout the years 1968-1971.

Aerodynamic Moments

We take the expression for the forces of aerodynamic pressure affecting an elementary area of the satellite's surface in the form (1.5) from Ref.(2):

$$\vec{f}_{\Sigma} = -\rho V^2\{(1-\varepsilon)(\vec{e}_V\vec{n})\vec{e}_V + 2\varepsilon(\vec{e}_V\vec{n})^2\vec{n} + (1-\varepsilon)\frac{v}{V}(\vec{e}_V\vec{n})\vec{n}\}ds \quad . \tag{1}$$

Here ε is the portion of molecules undergoing mirror reflection from the Sputnik surface whereas the rest appear to be reflected diffusely; v is the parameter proportional to the mean thermal velocity of molecules reflected diffusely; $\vec{e}_V = \vec{V}/V$; $\vec{n}$ is the vector of the normal to the satellite's surface. In contrast with Ref.(2) we shall assume $\vec{V}$ in this formula to be the total velocity of an elementary area relative to the atmosphere

$$\vec{V} = \vec{V}_0 + \Omega \times \vec{r}_c \quad , \tag{2}$$

where $\vec{V}_0$ is the satellite's orbital velocity (atmosphere rotation is not taken into account), $\vec{\Omega}$ is the vector of instantaneous angular velocity of the satellite's rotation, $\vec{r}_c$ is the radius vector drawn from the satellite's mass center.

The formula for the aerodynamic moment affecting the satellite is

$$\vec{M}_c = - \int_{s^*} \left[\vec{f}_\Sigma \times \vec{r}_c\right] ds$$

$$= \int_{s^*} \rho V^2 (\vec{e}_V \vec{n}) \{ (1-\varepsilon) \vec{e}_V$$

$$+ 2\varepsilon (\vec{e}_V \vec{n}) \vec{n} + (1-\varepsilon) \frac{v}{V} \vec{n}\} \times \vec{r}_c ds \qquad (3)$$

where s^* is a part of the satellite's surface interacting with the air flux (determined by a condition $(\vec{e}_V \cdot \vec{n}) > 0$).

We should bear in mind that irrespective of comparatively small sizes of a satellite the air density ρ, generally speaking, is not the same at different altitudes even within the satellite's length. As usual we shall assume in the first approximation that ρ varies according to the exponential law

$$\rho = \rho_0 e^{-\Delta h/H} \cong \rho_0 \left(1 - \frac{\Delta h}{H}\right) \quad ,$$

where H is the altitude of the homogeneous atmosphere, and ρ_0 corresponds to the altitude of the satellite's mass center. (The variation of ρ_0 with altitude during the satellite's orbiting of the Earth will be considered below.)

If we introduce $\vec{e}_r$ of unit geocentric radius vector of the satellite, Δh will be presented in the form $\Delta h = (\vec{e}_r \cdot \vec{r}_c)$ (see Fig. 1). Thus

$$\rho = \rho_0 \left[1 - \frac{(\vec{e}_r \vec{r}_c)}{H}\right] \quad . \qquad (4)$$

Then

$$\vec{M}_c = \rho_0 \int_{s^*} V^2(\vec{e}_V\vec{n})\{(1-\varepsilon)\vec{e}_V + 2\varepsilon(\vec{e}_V\vec{n})\vec{n} + (1-\varepsilon)\frac{v}{V}\vec{n}\}$$

$$\times \vec{r}_c ds + \left(-\frac{\rho_0}{H}\right)\int_{s^*} V^2(\vec{e}_V\vec{n})(\vec{e}_r\vec{r}_c)$$

$$\times \{(1-\varepsilon)\vec{e}_V + 2\varepsilon(\vec{e}_V\vec{n})\vec{n} + (1-\varepsilon)\frac{v}{V}\vec{n}\}\times\vec{r}_c ds \quad . \tag{5}$$

The second integral in this expression allows for the influence of air density variation within the length of the satellite; such an effect will be called an "aerogradient" effect.

The formula Eq. (5) is general, however; it is unsuitable in this form for obtaining particular results.

Consider the following data on the shape and mechanical parameters of a satellite: the satellite has axes of geometric and dynamic symmetry which do not coincide and at an angle κ to each other; the satellite has a plane of geometric symmetry which is perpendicular to the axis of geometric symmetry. Such a satellite will be called bisymmetrical. The reciprocal disposition of unit vectors of axes of geometric and dynamic symmetry $\vec{k}$ and $\vec{k}'$, geometric center O and mass center O' is determined by the distances R and $\Delta\ell$ and the angles κ and ζ as shown in Fig. 2. In this figure $\vec{i}'$, $\vec{j}'$, $\vec{k}'$ are unit vectors with the principal axes of the satellite's inertia ellipsoid in the direction $\vec{R}\perp\vec{k}$.

The vector $\vec{r}_c$ in the expression Eq. (5) can be presented in the form

$$\vec{r}_c = \vec{R} + \vec{k}\Delta\ell + \vec{r} \tag{6}$$

where $\vec{r}$ is the radius vector of a point on the satellite's surface drawn from its geometric center (see Fig. 2). We shall consider $R = |\vec{R}|$ and $\Delta\ell$ to be values of the first order of smallness compared with the mean value of $|\vec{r}|$.

Let us determine the vectors $\vec{k}$, $\vec{R}$ in terms of the unit vectors $\vec{i}'$, $\vec{j}'$, $\vec{k}'$. From Fig. 2 we find that

$$\vec{k} = \vec{i}'\sin\kappa + \vec{k}'\cos\kappa \quad , \tag{7}$$

$$\vec{R} = \frac{R}{\sqrt{1 - \sin^2\kappa\cos^2\zeta}} (\vec{i}'\cos\zeta + \vec{j}'\sin\zeta - \vec{k}\ \sin\kappa\cos\zeta) \quad , \tag{8}$$

or taking into account Eq. (7)

$$\vec{R} = \frac{R}{\sqrt{1 - \sin^2\kappa\cos^2\zeta}}$$

$$\times\ (\vec{i}'\cos\zeta\cos^2\kappa + \vec{j}'\sin\zeta - \vec{k}'\cos\zeta\sin\kappa\cos\kappa)$$

$$= \vec{i}'R_{x'} + \vec{j}'R_{y'} + \vec{k}'R_{z'} \quad . \tag{8'}$$

It should be noted that the correction for rotational motion $\vec{\Omega}\times\vec{r}_c$ from Eq. (2) is a value of the first order of smallness compared with the orbital velocity of the satellite $\vec{V}_0$. Substituting Eq. (2) into Eq. (5) and neglecting the terms of the second order of smallness we obtain

$$\vec{M}_c = \vec{M}_1 + \vec{M}_2 + \vec{M}_3 \quad , \tag{9}$$

where

$$\vec{M}_1 = \rho_0 V_0^2 \int_{s^*} (\vec{e}_V \vec{n}) \{ (1-\varepsilon) \vec{e}_V + 2\varepsilon (\vec{e}_V \vec{n}) \vec{n}$$

$$+ (1-\varepsilon) \frac{v}{V} \vec{n} \} \times \vec{r}_c ds \quad , \tag{10}$$

$$\vec{M}_2 = \rho_0 V_0 (1-\varepsilon) \int_{s^*} \{ (\vec{e}_V \vec{r}_c) \left[(\vec{\Omega} \times \vec{r}_c) \vec{n} \right]$$

$$+ (\vec{e}_V \vec{n}) (\vec{\Omega} \times \vec{r}_c) \times \vec{r}_c \} ds$$

$$+ 4 \rho_0 V_0 \varepsilon \int_{s^*} (\vec{e}_V \vec{n}) \left[(\vec{\Omega} \times \vec{r}_c) \vec{n} \right] (\vec{n} \times \vec{r}_c) ds$$

$$+ \rho_0 (1-\varepsilon) v \int_{s^*} \left[(\vec{\Omega} \times \vec{r}_c) \vec{n} \right] (\vec{n} \times \vec{r}_c) ds \quad , \tag{11}$$

$$\vec{M}_3 = \frac{\rho_0 V_0^2}{H} \int_{s^*} (\vec{e}_V \vec{n}) (\vec{e}_r \vec{r}_c) \{ (1-\varepsilon) \vec{e}_V + 2 (\vec{e}_V \vec{n}) \vec{n}$$

$$+ (1-\varepsilon) \frac{v}{V} \vec{n} \} \times \vec{r}_c ds \quad . \tag{12}$$

$\vec{e}_V$ is now used in the sense of $e_V = \vec{V}_0 / V_0$.

For simplicity we shall write ρ_0, V_0 henceforth as ρ, V.

As is seen from Eqs. (10), (11) and (12) $\vec{M}_1$ is a moment produced by the forces of aerodynamic pressure on the non-rotating satellite, $\vec{M}_2$ is the proportional to the absolute value of angular velocity Ω, $\vec{M}_3$ is proportional to the atmosphere density gradient 1/H. Let us call the moments $\vec{M}_1$, $\vec{M}_2$, $\vec{M}_3$ the moments of aerodynamic pressure, aerodynamic friction and aerogradient respectively. (In the case of a small satellite length the influence of moment $\vec{M}_3$ must be quite small; moreover, the aerogradient effect will be decreased because of density fluctuations associated with local fluctuations in the atmosphere estimates. Therefore estimates of the effect given in the section ahead in this text on moment $\vec{M}_3$, are maximum estimates.)

Substituting Eq. (6) into Eqs. (10), (11) and (12) and neglecting terms of the second order of smallness we find that

$$\vec{M}_1 = \vec{F}\times\vec{R} + \vec{F}\times\vec{k}\Delta\ell + \vec{M}_0 \quad , \tag{13}$$

where

$$\vec{F} = \rho V^2 \int_{s^*} (\vec{e}_V\vec{n})\{(1-\varepsilon)\vec{e}_V + 2(\vec{e}_V\vec{n})\vec{n} + (1-\varepsilon)\frac{v}{V}\vec{n}\}ds \quad . \tag{14}$$

The moments $\vec{M}_0$, $\vec{M}_2$, $\vec{M}_3$ are now expressed by formulae (10), (11) and (12) respectively in which $\vec{r}$ appears instead of $\vec{r}_c$. We will refer to this new version as (10'), (11') and (12').

Consider independently the aerodynamic moments of Eq. (9) for the specific case of a bisymmetrical satellite. For this we introduce a right-handed rectangular coordinate system xyz with unit vectors $\vec{i}$, $\vec{j}$, $\vec{k}$ (Fig. 3). The system is related to the satellite as follows: the origin of the coordinate system coincides with the geometric center of the satellite and the axis z with that of geometric symmetry so that $\vec{OA} = -\vec{k}\Delta\ell$ (Fig. 2), the direction of the y axis is such that the vector $\vec{e}_V$ lies in the plane yz, and $(\vec{j}\cdot\vec{e}_V)>0$: let the angle between $\vec{e}_V$ and $\vec{k}$ be designated by δ. We may call this system of coordinates a semiconnected one, the vectors $\vec{e}_V$, $\vec{n}$, $\vec{r}$ in this system become

$$\vec{e}_V = \vec{i}\sin\delta + \vec{k}\cos\delta$$

$$\vec{n} = \vec{i}n_x + \vec{j}n_y + \vec{k}n_z$$

$$\vec{r} = \vec{i}x + \vec{j}y + \vec{k}z \quad . \tag{15}$$

Moment $\vec{M}_1$

Consider the aerodynamic pressure force moment $\vec{M}_1$. Insert Eq. (15) into (10') and integrate (10') over the surface. The formula (10') for an axisymmetric satellite is completely symmetrical with respect to the plane yz. Therefore the expressions containing the factor x or n_x become zero on integration:

$$\int_{s^*} x\, f(y,z,n_y,n_z,\delta)ds = 0 \quad ;$$

$$\int_{s^*} n_x f(y,z,n_y,n_z,\delta)ds = 0 \quad , \tag{16}$$

where f is an arbitrary function of its arguments.

It may be also shown (see Ref. 3) that for the bisymmetrical satellite

$$\int_{s^*} n_z y f(x,n_x,\delta)ds = 0 \quad ;$$

$$\int_{s^*} n_y z f(x,n_x,\delta)ds = 0 \quad ;$$

$$\int_{s^*} (n_z z - n_y y) f(x, n_x, \delta) ds = 0 \quad ; \tag{17}$$

where f is an arbitrary function of its arguments.

Integrating by taking into account Eq. (16) and (17) we obtain

$$\vec{M}_0 = \rho V^2 C(\delta) (\vec{e}_V \times \vec{k}) \quad , \tag{18}$$

where

$$C(\delta) = 2\varepsilon \int_{s^*} (\vec{e}_V \vec{n})^2 (n_y z - n_z y) \frac{ds}{\sin\delta}$$

$$+ (1-\varepsilon) \frac{v}{V} \int_{s^*} (\vec{e}_V \vec{n}) (n_y z - n_z y) \frac{ds}{\sin\delta} \quad . \tag{19}$$

It should be noted that in the expression (19) for $C(\delta)$ there are coefficients ε and v/V (in the first and second terms respectively). Thus at a completed (inelastic collision of molecules with the surface of a bisymmetrical satellite

$$C(\delta) = 0 \quad . \tag{20}$$

Taking $\varepsilon = 0.02$, $v/V = 0.1$ as in Ref. (2) we may consider that the Eq. (20) holds true approximately for an actual satellite. As for a cylindrical satellite the Eq. (20) can be considered exact.

Let us now calculate $\vec{F}$. With this end in view we represent the vector $\vec{n}$ in the form

$$\vec{n} = n_z \vec{i} + \frac{n_y}{\sin\delta} \vec{e}_V + (n_z - \text{ctg}\,\delta \cdot n_y) \vec{k} \quad . \tag{21}$$

Insert $\vec{n}$ from Eq. (21) and $\vec{e}_V$ from Eq. (15) into (14). Integrating and allowing for Eq. (16) we derive

$$\vec{F} = \rho V^2 \left[F_V(\delta)\vec{e}_V + F_k(\delta)\vec{k} \right] \quad , \tag{22}$$

where

$$F_V(\delta) = (1-\varepsilon) \int_{s^*} (\vec{e}_V\vec{n})\,ds + 2\varepsilon \int_{s^*} (\vec{e}_V\vec{n})^2 \frac{n_y}{\sin\delta}\,ds$$

$$+ (1-\varepsilon)\,\frac{v}{V} \int_{s^*} (\vec{e}_V\vec{n}) \frac{n_y}{\sin\delta}\,ds \quad , \tag{23}$$

$$F_k(\delta) = 2\varepsilon \int (\vec{e}_V\vec{n})^2 (n_z - n_y \mathrm{ctg}\delta)\,ds$$

$$+ (1-\varepsilon)\,\frac{v}{V} \int_{s^*} (\vec{e}_V\vec{n})(n_z - n_y \mathrm{ctg}\delta)\,ds \quad . \tag{24}$$

Thus

$$\vec{M}_1 = \rho V^2 \{ F_V(\delta)(\vec{e}_V \times \vec{R}) + F_k(\delta)(\vec{k}\times\vec{R})$$

$$+ \left[F_V(\delta)\Delta\ell + C(\delta) \right] (\vec{e}_V \times \vec{k}) \Big] = \vec{M}_{11} + \vec{M}_{12} + \vec{M}_{13} \quad . \tag{25}$$

Let us clarify whether the moment of aerodynamic pressure after time averaging has an accelerating component, i.e. a component parallel to the vector of the kinetic moment $\vec{L}$. (Here "acceleration" is used to imply both positive and negative acceleration.) For this purpose it is necessary to know the form of the functions $F_V(\delta)$ and $F_k(\delta)$. (As mentioned above $C(\delta)$ is about zero, and as is shown from the formula (25) $C(\delta)$ in no case produces a qualitatively new effect in the satellite's motion: the moment $\vec{M}_{13}$ exists whichever of the two terms is present.)

The form of the functions $F_V(\delta)$ and $F_k(\delta)$ depends on the shape of the satellite (it should be noted that generally speaking $s^* = s^*(\delta)$). According to the papers (2,12) the following approximations for the functions $F_V(\delta)$ and $F_k(\delta)$ will be taken

$$F_V(\delta) = F_{V_0} + F_{V_1}\cos\delta + F_{V_2}\cos^2\delta \quad , \tag{26}$$

$$F_k(\delta) = F_{k_0} + F_{k_1}\cos\delta + F_{V_2}\cos^2\delta \quad . \tag{27}$$

Let us project the moment $\vec{M}_1$ on to the direction of the satellite's angular moment vector and average the projection over time. In this case the following systems of coordinates should be introduced.

1. The perigee system of coordinates XYZ with unit vectors $\vec{e}_X$, $\vec{e}_Y$, $\vec{e}_Z$ (see Fig. 4). The axis X is along the velocity vector at the orbit perigee, the axis Z is along the radius vector of the orbit perigee of the satellite, the axis Y is along the normal to the plane of the satellite's orbit.
2. The system of coordinates $L_1L_2L_3$ connected with the angular moment $\vec{L}$ (Figs. 4 and 5). The axis L_3 is along the vector of the angular moment $\vec{L}$; the axis L_2 lies along the straight line of intersection of the plane perpendicular to L_3 with the plane XZ (see Fig. 4); the direction of the axis L_1 is such as to make the system of coordinates a right-handed one. The position of the system $L_1L_2L_3$ in respect to XYZ is determined by the angles Θ and λ as is shown in Fig. 4.
3. The system of coordinates x'y'z' with unit vectors $\vec{e}_{x'}$, $\vec{e}_{y'}$, $\vec{e}_{z'}$ rigidly connected with the inertia ellipsoid of the satellite (Fig. 5 and Fig. 2). The axes x', y', z' are along the major axes of this ellipsoid, the axis z' being directed along the direction of that of smallest moment of inertia. Insofar as the satellite is dynamically symmetrical, the choice of axis x' in the plane perpendicular to axis z' is arbitrary. Let us determine

the rotation of the x'y'z' system relative to the $L_1L_2L_3$ system by Euler's angles θ, ψ, ϕ as is shown in Fig. 5. Transformation matrices are involved between the above systems of coordinates

	x'	y'	z'
X	α_1	α_2	α_3
Y	β_1	β_2	β_3
Z	γ_1	γ_2	γ_3

(28)

	L_1	L_2	L_3
X	m_1	m_2	m_3
Y	n_1	n_2	n_3
Z	ℓ_1	ℓ_2	ℓ_3

(29)

	x'	y'	z'
L_1	α_{11}	α_{12}	α_{13}
L_2	α_{21}	α_{22}	α_{23}
L_3	α_{31}	α_{32}	α_{33}

(30)

It follows from Fig. 4 that

$$m_1 = -\sin\Theta \;; \quad m_2 = 0 \;; \quad m_3 = \cos\Theta \;;$$
$$n_1 = -\cos\Theta\sin\lambda; \quad n_2 = -\cos\lambda; \quad n_3 = -\sin\Theta\sin\lambda;$$
$$\ell_1 = \cos\Theta\cos\lambda \;; \quad \ell_2 = -\sin\lambda; \quad \ell_3 = \sin\Theta\cos\lambda \;; \tag{31}$$

and from Fig. 5 we derive an explicit series of coefficients α_{ij}:

$$\alpha_{11} = \cos\psi\cos\phi - \sin\psi\sin\phi\cos\theta \;;$$

$$\alpha_{12} = -\cos\psi\sin\phi - \sin\psi\cos\phi\cos\theta \;;$$

$$\alpha_{13} = \sin\theta\sin\psi \;;$$

$$\alpha_{21} = \sin\psi\cos\phi + \cos\psi\sin\phi\cos\theta \;;$$

$$\alpha_{22} = -\sin\psi\sin\phi + \cos\psi\cos\phi\cos\theta \;;$$

$$\alpha_{23} = -\sin\theta\cos\psi \quad ;$$

$$\alpha_{31} = \sin\theta\sin\phi \quad ;$$

$$\alpha_{32} = \sin\theta\cos\phi \quad ;$$

$$\alpha_{33} = \cos\theta \quad . \qquad (32)$$

Since the work of the perturbation forces is small compared with the kinetic energy of the satellite motion around the mass center, this motion in the first approximation can be considered as a non-perturbed one. In this case the angular moment $\vec{L}$ vector has a constant modulus and a constant direction in space; hence it follows Θ = const, λ = const. In addition, the non-perturbed motion is defined by three more constant parameters θ, $\dot{\psi}$ and $\dot{\phi}$. (The angular velocity of spin rotation $\dot{\phi}$ is not an independent value and is associated with the precession velocity $\dot{\psi}$ by relationship (84), see below).

Because of these perturbations the above five parameters will vary with time and their variation will describe the perturbed motion (5). As far as the marked spin-rate variation proceeds within the order of one or more days, the variables ψ, ϕ, ν can be regarded as high-speed ones. Cases of resonance between the changes in the velocity of these variables are not considered here, therefore the time averaging of moments can be replaced by three independent averagings over the variables ψ, ϕ, ν.

The vector of the satellite orbital velocity ($\vec{e}_V$) is a function of a true anomaly ν of the satellite. Note that ν is not a linear function of time, therefore the averaging of a value B over ν is of the form

$$\bar{B}^{\nu} = \frac{1}{P}\int_0^P B dt = \frac{1}{2\pi}\int_0^{2\pi} B \frac{dt}{d\nu} d\nu = \frac{1}{2\pi}\int_0^{2\pi} B \frac{(1-e^2)^{3/2} d\nu}{(1+e\ \cos\nu)^2} . \tag{33}$$

So let us start with the averaging of terms in the right-hand part of Eq. (25). We put down the expressions $\cos\delta$ $\cos^2\delta$ in the coordinate form. According to Fig. 3

$$\cos\delta = (\vec{e}_V \cdot \vec{k}) \quad .$$

In the perigee system of coordinates XYZ, the vector $\vec{e}_V$ is written as follows

$$\vec{e}_V = \vec{e}_X \frac{e+\cos\nu}{\sqrt{1+e^2+2e\ \cos\nu}} + \vec{e}_Z \frac{-\sin\nu}{\sqrt{1+e^2+2e\ \cos\nu}} \tag{34}$$

where e is the orbital eccentricity of the satellite and ν is the satellite's true anomaly.

The vector $\vec{k}$ in the x'y'z' coordinate system is presented by formula (7).

The vectors $\vec{e}_V$ and $\vec{k}$ will be given through Eqs. (29) and (30) in the coordinate system $L_1L_2L_3$

$$\vec{e}_V = \vec{e}_{L_1} e_{V_1}(\nu,\Theta,\lambda) + \vec{e}_{L_2} e_{V_2}(\nu,\Theta,\lambda) + \vec{e}_{L_3} e_{V_3}(\nu,\Theta,\lambda); \tag{35}$$

$$\vec{k} = \vec{e}_{L_1} k_1(\psi,\phi,\theta) + \vec{e}_{L_2} k_2(\psi,\phi,\theta) + \vec{e}_{L_3} k_3(\psi,\phi,\theta) \quad , \tag{36}$$

where

$$e_{V_1} = \frac{1}{\sqrt{1+e^2+2e\ \cos\nu}} \left[m_1(e+\cos\nu)+\ell_1(-\sin\nu)\right] \quad ;$$

$$e_{V_2} = \frac{1}{\sqrt{1+e^2+2e\cos\nu}} \left[m_2(e+\cos\nu)+\ell_2(-\sin\nu)\right] \quad ;$$

$$e_{V_3} = \frac{1}{\sqrt{1+e^2+2e\cos\nu}} \left[m_3(e+\cos\nu)+\ell_3(-\sin\nu)\right] \quad ;$$

$$(37)$$

$$k_1 = \sin\kappa\alpha_{11} + \cos\kappa\alpha_{13}$$

$$k_2 = \sin\kappa\alpha_{21} + \cos\kappa\alpha_{23}$$

$$k_3 = \sin\kappa\alpha_{31} + \cos\kappa\alpha_{33} \quad .$$

Then

$$\cos\delta = (\vec{e}_V \cdot \vec{k}) = \sum_{i=1}^{3} e_{V_i}(\nu,\Theta,\lambda)k_i(\psi,\phi,\theta) \quad , \quad (38)$$

$$\cos^2\delta = \sum_{\substack{i=1\\j=1}}^{3} e_{V_i}(\nu,\Theta,\lambda)e_{V_j}(\nu,\Theta,\lambda)k_i(\psi,\phi,\theta)k_j(\psi,\phi,\theta) \; . \quad (39)$$

When considering the averaging of the variables ψ, ϕ, ν we shall write $e_{V_i}(\nu)$, $k_i(\psi,\phi)$ from now on for brevity. According to Eq. (25) and (26) $\vec{M}_{11}$ can be expressed as

$$\vec{M}_{11} = \rho V^2 F_V(\delta)(\vec{e}_V \times \vec{R})$$

$$= \rho V^2 (F_{V_0} + F_{V_1}\cos\delta + F_{V_2}\cos^2\delta)(\vec{e}_V \times \vec{R}) \quad . \quad (40)$$

In Eq. (40) V is satellite's orbital velocity, ρ is the atmosphere density in the given point of the orbit. It is shown in (1) that

$$V^2(\nu) = \frac{\mu}{P}\bar{V}^2(\nu) \quad , \quad (41)$$

$$\rho(\nu) = \rho_\pi \cdot \bar{\rho}(\nu) \quad , \quad (42)$$

where

$$\left.\begin{aligned}\bar{V}^2(\nu) &= 1+e^2+2e\cos\nu \\ \bar{\rho}(\nu) &= \exp\left[-\frac{1}{H}\frac{pe}{1+e}\frac{1-\cos\nu}{1+e\cos\nu}\right]\end{aligned}\right\} \tag{43}$$

μ is the Earth's gravitational constant; p is a focal parameter of the satellite's orbit; e is the satellite's orbital eccentricity; ν is the satellite's true anomaly; H is the density scale height; ρ_π is the atmosphere density in the orbit perigee.

Applying Eq. (8') and (30) we can present the vector $\vec{R}$ in the coordinate system $L_1L_2L_3$ as follows

$$\begin{aligned}\vec{R} &= \vec{e}_{L_1}(R_{x'}\alpha_{11}+R_{y'}\alpha_{12}+R_{z'}\alpha_{13}) \\ &+ \vec{e}_{L_2}(R_{x'}\alpha_{21}+R_{y'}\alpha_{22}+R_{z'}\alpha_{23}) \\ &+ \vec{e}_{L_3}(R_{x'}\alpha_{31}+R_{y'}\alpha_{32}+R_{z'}\alpha_{33}) \\ &= \vec{e}_{L_1}R_{L_1} + \vec{e}_{L_2}R_{L_2} + \vec{e}_{L_3}R_{L_3} \quad .\end{aligned} \tag{44}$$

The $\vec{M}_{11}$ moment projection on the vector of an angular moment $\vec{L}$ is of the form

$$\begin{aligned}(\vec{M}_{11})_{\vec{L}} &= \left[\rho V^2 F_V(\delta)(\vec{e}_V \times \vec{R})\right]_{\vec{L}} \\ &= \rho V^2 F_V(\delta)(e_{V_1}R_{L_2} - e_{V_2}R_{L_1}) \quad .\end{aligned} \tag{45}$$

To express Eq. (45) in terms of the variables ψ, ϕ, ν we use Eqs. (26, 37-39, 41-44) and average the expression obtained over the variables ψ, ϕ, ν obtaining

$$\overline{(\vec{M}_{11})_{\vec{L}}}^{\psi,\phi,\nu} = \rho_{\pi} \frac{\mu}{p} R_{Y'} \sin\kappa$$

$$\times \left\{ F_{V_1} \frac{1}{2} \cos\theta \left(\overline{\bar{\rho}\bar{V}^2 e_{V_1}^2}^{\nu} + \overline{\bar{\rho}\bar{V}^2 e_{V_2}^2}^{\nu} \right) \right.$$

$$+ F_{V_2} \cos\kappa \left(\cos\theta - \frac{1}{2}\sin^2\theta\right)$$

$$\left. \times \left(\overline{\bar{\rho}\bar{V}^2 e_{V_1}^2 e_{V_3}}^{\nu} + \overline{\bar{\rho}\bar{V}^2 e_{V_2}^2 e_{V_3}}^{\nu} \right) \right\} , \quad (46)$$

where

$$R_{Y'} = \frac{R\sin\zeta}{\sqrt{1-\sin^2\kappa\cos^2\zeta}}$$

(see Fig. 2 and Eq. (8');

$$\overline{\bar{\rho}\bar{V}^2 e_{V_1}^2}^{\nu} = m_1^2 I_1 + \ell_1^2 I_2 ,$$

$$\overline{\bar{\rho}\bar{V}^2 e_{V_2}^2}^{\nu} = m_2^2 I_1 + \ell_2^2 I_2 ,$$

$$\overline{\bar{\rho}\bar{V}^2 e_{V_1}^2 e_{V_3}}^{\nu} = m_1^2 m_3 J_1 + (\ell_1^2 m_3 + 2m_1 \ell_1 \ell_3) J_2 ;$$

$$\overline{\bar{\rho}\bar{V}^2 e_{V_2}^2 e_{V_3}}^{\nu} = m_2^2 m_3 J_1 + (\ell_2^2 m_3 + 2m_2 \ell_2 \ell_3) J_2 ; \quad (47)$$

$$I_1 = \frac{(1-e^2)^{3/2}}{2\pi} \int_0^{2\pi} \frac{\bar{\rho}(\nu)(e+\cos\nu)^2}{(1+e\cos\nu)^2} d\nu ,$$

$$I_2 = \frac{(1-e^2)^{3/2}}{2\pi} \int_0^{2\pi} \frac{\bar{\rho}(\nu)\sin^2\nu d\nu}{(1+e\cos\nu)^2} ;$$

$$J_1 = \frac{(1-e^2)^{3/2}}{2\pi} \int_0^{2\pi} \frac{\bar{\rho}(\nu)(e+\cos\nu)^3 d\nu}{(1+e\cos\nu)^2 \sqrt{1+e^2+2e\cos\nu}} \quad ;$$

$$J_2 = \frac{(1-e^2)^{3/2}}{2\pi} \int_0^{2\pi} \frac{\bar{\rho}(\nu)(e+\cos\nu)\sin^2\nu d\nu}{(1+e\cos\nu)^2 \sqrt{1+e^2+2e\cos\nu}} \quad ; \tag{48}$$

The third term of Eq. (25) is of the form

$$\vec{M}_{13} = \rho V^2 \left[F_V(\delta)\Delta\ell + C(\delta) \right] (\vec{e}_V \times \vec{k}) \quad .$$

Taking into account Eq. (20) we obtain

$$\vec{M}_{13} = \rho V^2 (F_{V_0} + F_{V_1}\cos\delta + F_{V_2}\cos^2\delta)(\vec{e}_V \times \vec{k}) \quad . \tag{49}$$

Once again we have to repeat that if $C(\delta) \neq 0$ it will not change the qualitative picture of motion, just the factors F_{V_0}, F_{V_1}, F_{V_2} will be somewhat changed in Eq. (49).

After projecting Eq. (49) on the vector of the satellite's angular moment $\vec{L}$ we obtain

$$(\vec{M}_{13})_{\vec{L}} = \rho V^2 (F_{V_0} + F_{V_1}\cos\delta + F_{V_2}\cos^2\delta)(e_{V_1}k_2 - e_{V_2}k_1) . \tag{50}$$

Averaging over variables ψ, ϕ, ν (Eq. 50) we get

$$\overline{(\vec{M}_{13})_{\vec{L}}}^{\psi,\phi,\nu} = 0 \quad . \tag{51}$$

We average now the second term of Eq. (25). Applying formula (8) we obtain

$$\vec{M}_{12} = \rho V^2 F_k(\delta)(\vec{k} \times \vec{R}) = \frac{\rho V^2 F_k(\delta) R}{\sqrt{1-\sin^2\kappa\cos^2\zeta}} \times \left[\vec{k} \times (\vec{i}\,'\cos\zeta + \vec{j}\sin\zeta) \right] \quad . \tag{52}$$

Let us insert Eq. (7) into Eq. (52). Then

$$\vec{M}_{12} = \frac{\rho V^2 F_k(\delta) R}{\sqrt{1-\sin^2\kappa\cos^2\zeta}} \left[\vec{i}'(-\cos\kappa\sin\zeta) + \vec{j}'(\cos\kappa\cos\zeta) + \vec{k}'(\sin\kappa\sin\zeta)\right] . \quad (53)$$

Project (53) on to the vector of angular moment $\vec{L}$ by finding the projections of unit vectors $\vec{i}'$, $\vec{j}'$, $\vec{k}'$ on the unit vector $\vec{e}_{L_3}$ through transformation matrix (30);

$$(\vec{M}_{12})_{\vec{L}} = \frac{\rho V^2 F_k(\delta) R}{\sqrt{1-\sin^2\kappa\cos^2\zeta}} \left[-\alpha_{31}\cos\kappa\sin\zeta + \alpha_{32}\cos\kappa\cos\zeta + \alpha_{33}\sin\kappa\sin\zeta\right] . \quad (54)$$

Then approximation (27), expanded expressions (38) and (39) for $\cos\delta$ and $\cos^2\delta$ are inserted into Eq. (54), and the averaging over ψ, ϕ, ν is made.

As a result we have

$$\overline{(\vec{M}_{12})_{\vec{L}}}^{\psi,\phi,\nu} = \rho_\pi \frac{\mu}{p} R_{y'} \sin\kappa \Big\{ F_{k_0}\cos\theta G_1 + F_{k_1}\cos\kappa\left(1-\frac{3}{2}\sin^2\theta\right) m_3 G_2 + F_{k_2}\cos\theta \times \Big[\cos^2\kappa\sin^2\theta \cdot \frac{1}{2}\left((m_1^2+m_2^2-2m_3^2) I_1 + (\ell_1^2+\ell_2^2-2\ell_3^2) I_2\right) + \left\{\frac{1}{4}\sin^2\kappa(1+\cos^2\theta) + \frac{1}{2}\cos^2\kappa\sin^2\theta\right\}\left((m_1^2+m_2^2) I_1 + (\ell_1^2+\ell_2^2) I_2\right) + \left\{\frac{1}{2}\sin^2\kappa\sin^2\theta + \cos^2\kappa\cos^2\theta\right\}(m_3^2 I_1 + \ell_3^2 I_2)\Big]\Big\} , \quad (55)$$

where

$$G_1 = \frac{(1-e^2)^{3/2}}{2\pi} \int_0^{2\pi} \bar{\rho}(\nu) \frac{(1+e^2+2e\cos\nu)}{(1+e\cos\nu)^2} d\nu \quad ;$$

$$G_2 = \frac{(1-e^2)^{3/2}}{2\pi} \int_0^{2\pi} \bar{\rho}(\nu) \frac{\sqrt{1+e^2+2e\cos\nu}\,(e+\cos\nu)}{(1+e\cos\nu)^2} d\nu \, . \tag{56}$$

Thus formulae are derived for the accelerating components of the moment $\vec{M}_1$. As is seen from Eqs. (46), (51) and (55) the value of the angular moment is affected by terms of the form $\vec{e}_V \times \vec{R}$ and $\vec{k} \times \vec{R}$ from Eq. (25).

It should be noted that in Eq. (23) for $F_V(\delta)$ the principal term is conditioned by the absolutely inelastic collision of molecules with the satellite's surface, and in Eq. (24) for $F_k(\delta)$ the first and second terms are caused by the pressure of molecules reflected directly or diffusely. Let us call the moments described by $\vec{M}_{11}$ and $\vec{M}_{12}$ (first and second terms of Eq. (25)) the moments of completely inelastic auto-rotation and of diffuse-mirror auto-rotation respectively.

The term of form $\vec{e}_V \times \vec{k}$ causes no change in the value of the angular moment: we call it a conservative moment (2). Thus the moment $\vec{M}_1$ has an accelerating component in the case $R_{y'} \neq 0$, i.e. if the axes of geometric and dynamic symmetry have no common point (see Fig. 2).

Let us calculate the coefficients F_{V_0}, F_{V_1}, F_{V_2}, F_{k_0}, F_{k_1}, F_{k_2} for the satellite of a right circular cylinder-shape with flat butt-ends of radius R_0 and of length ℓ.

Integrating Eqs. (23), (24) for the case of a right circular cylinder we find

$$F_V(\delta) = R_0 \ell \left[\left(2 + \frac{2}{3}\varepsilon \right) \sin\delta + \frac{\pi}{2} \frac{(1-\varepsilon)v}{V} + \frac{(1-\varepsilon)\pi R_0}{\ell} |\cos\delta| \right] , \quad (57)$$

$$F_k(\delta) = \cos\delta \left\{ \frac{(1-\varepsilon)v}{V} \pi R_0 \left(R_0 - \frac{\ell}{2} \right) + 2\varepsilon R_0 \left[\pi R_0 |\cos\delta| - \frac{4}{3} \ell \sin\delta \right] \right\} . \quad (58)$$

Let us take the following approximations (2)

$$\sin\delta \cong \sin^2\delta \; ; \quad |\cos\delta| \cong \cos^2\delta \; ;$$

$$\cos\delta |\cos\delta| \cong \cos\delta \; . \quad (59)$$

In this case the terms $F_V(\delta)$ and $F_k(\delta)$ will be of the form

$$F_{V_0} = R_0 \ell \left[\left(2 + \frac{2}{3}\varepsilon \right) + \frac{\pi}{2} \frac{(1-\varepsilon)v}{V} \right] \; ;$$

$$F_{V_1} = 0 \; ;$$

$$F_{V_2} = R_0 \ell \left[(1-\varepsilon) \frac{\pi R_0}{\ell} - \left(2 + \frac{2}{3}\varepsilon \right) \right] \; ;$$

$$F_{k_0} = 0 \; ;$$

$$F_{k_1} = \frac{(1-\varepsilon)v}{V} \pi R_0 \left(R_0 - \frac{\ell}{2} \right) + 2\varepsilon R_0 \left(\pi R_0 - \frac{4}{3} \ell \right) \; ;$$

$$F_{k_2} = 0 \; . \quad (60)$$

Inserting Eq. (59) into (58) we have obtained a term with $\cos^3\delta$ in the expression for $F_k(\delta)$ but since in Eq.

(27) there is no corresponding term, this term is not considered, i.e. the moment of diffusion-mirror autorotation is considered in approximation (27).

Numerical estimations and comparison data of all the moments will be summarized in the section on Estimation and Comparison of Effects of Moments.

Moment $\vec{M}_2$

Consider now the friction moment $\vec{M}_2$. Let us designate the projection of the vector of instantaneous angular velocity of the satellite's rotation $\vec{\Omega}$ on the axes of a semi-connected coordinate system (Fig. 3) by p, q, r respectively:

$$\vec{\Omega} = p\vec{i} + q\vec{j} + r\vec{k} \quad . \tag{61}$$

Inserting Eq. (61) into (11'), resolving the expression under the integral over the unit vectors $\vec{i}$, $\vec{j}$, $\vec{k}$ and taking into account Eq. (16) and the fact that $n_x y - n_y x = 0$ for the axisymmetric satellite we obtain the expression for $\vec{M}_2$ in the form

$$\vec{M}_2 = \rho V \{ \vec{i}[-I_5(\delta)p] + \vec{j}[-I_3(\delta)q + I_4(\delta)r] + \vec{k}[-I_1(\delta)r + I_2(\delta)q] \} \quad , \tag{62}$$

where

$$I_1(\delta) = (1-\varepsilon) \int_{s^*} (n_y \sin\delta + n_z \cos\delta)(x^2+y^2)\,ds \quad ;$$

$$I_2(\delta) = (1-\varepsilon) \int_{s^*} [x(xn_z - zn_x)\sin\delta + (n_y \sin\delta + n_z \cos\delta) yz]\,ds \quad ;$$

$$I_3(\delta) = (1-\varepsilon)\int_{S^*} [x(xn_z - zn_x)\cos\delta + (n_y \sin\delta + n_z\cos\delta)(x^2+z^2)]ds + 4\varepsilon \int_{S^*} (n_y\sin\delta + n_z\cos\delta) \times (n_z x - n_x z)^2 ds + (1-\varepsilon)\frac{v}{V}\int_{S^*} (n_z x - n_x z)^2 ds \quad ;$$

$$I_4(\delta) = (1-\varepsilon)\int_{S^*} (n_y\sin\delta + n_z\cos\delta) yz \, ds \quad ;$$

$$I_5(\delta) = (1-\varepsilon)\int_{S^*} [(y\cos\delta - z\sin\delta)(yn_z - zn_y) + (n_z\cos\delta + n_y\sin\delta)(z^2+y^2)ds + 4\varepsilon\int_{S^*} (n_y\sin\delta + n_z\cos\delta)(n_y z - yn_z)^2 ds + (1-\varepsilon)\frac{v}{V}\int_{S^*} (n_y z - yn_z)^2 ds \quad . \tag{63}$$

The components of the vector of instantaneous angular velocity $\vec{\Omega}$ are connected with those of the vector $\vec{L}$ as follows

$$p = \frac{L_x}{A} \quad ; \quad q = \frac{L_y}{A} \quad ; \quad r = \frac{L_z}{C} \quad , \tag{64}$$

where A and C are the satellite's moments of inertia relative to the transverse and longitudinal axes of the inertia ellipsoid, and

$$L_x = L(\vec{e}_{L_3}\cdot\vec{i}) \quad ; \quad L_y = L(\vec{e}_{L_3}\cdot\vec{j}) \quad ; \quad L_z = L(\vec{e}_{L_3}\cdot\vec{k}) \tag{65}$$

Projecting Eq. (62) on to the vector $\vec{L}$ and taking into consideration Eqs. (64) and (65) we obtain

$$(\vec{M}_2)_{\vec{L}} = (\vec{M}_2\vec{e}_{L_3}) = \rho VL \left\{ -\frac{I_5(\delta)}{A}(\vec{e}_{L_3}\cdot\vec{i})^2 - \frac{I_3(\delta)}{A}(\vec{e}_{L_3}\cdot\vec{j})^2 - \frac{I_1(\delta)}{C}(\vec{e}_{L_3}\cdot\vec{k})^2 + \left[\frac{I_2(\delta)}{A} + \frac{I_4(\delta)}{C}\right](\vec{e}_{L_3}\cdot\vec{j})(\vec{e}_{L_3}\cdot\vec{k}) \right\} . \quad (66)$$

Let us plot the traces of vectors $\vec{e}_{L_3}$, $\vec{e}_V$, $\vec{i}$, $\vec{j}$, $\vec{k}$ (Fig. 6) in the celestial sphere and introduce angles $\tilde{\theta}$ and β.

It is obvious that

$$(\vec{e}_{L_3}\cdot\vec{i}) = -\sin\tilde{\theta}\sin\beta \quad ; \quad (\vec{e}_{L_3}\cdot\vec{j}) = \sin\tilde{\theta}\cos\beta \quad ;$$
$$(\vec{e}_{L_3}\cdot\vec{k}) = \cos\tilde{\theta} \quad . \quad (67)$$

Inserting Eq. (67) into (66) we get

$$(\vec{M}_2)_{\vec{L}} = \rho VL \left\{ -\frac{I_5(\delta)}{A}\sin^2\tilde{\theta} - \frac{I_1(\delta)}{C}\cos^2\tilde{\theta} + \frac{I_5(\delta) - I_3(\delta)}{A}\sin^2\tilde{\theta}\cos^2\beta + \left\{\frac{I_2(\delta)}{A} + \frac{I_4(\delta)}{C}\right\}\cos\tilde{\theta}\sin\tilde{\theta}\cos\beta \right\} . \quad (68)$$

According to Fig. 6 we find that

$$\cos\beta = \frac{(\vec{e}_{L_3}\cdot\vec{e}_V) - \cos\tilde{\theta}\cos\delta}{\sin\tilde{\theta}\sin\delta} ,$$

or

$$\cos\beta\sin\delta = \frac{1}{\sin\tilde{\theta}}(e_{V_3} - \cos\tilde{\theta}\cos\delta) . \quad (69)$$

In a general case the coefficients $I_i(\delta)$ could be approximated by expansions of the form

$$I_i(\delta) = I_{i_0} + I_{i_1}\cos\delta + \ldots + I_{i_n}\cos^n\delta \quad . \qquad (70)$$

However, at the insertion of Eq. (70) into (68) there appear expressions containing $\sin\delta$ or $\sin^2\delta$ in the denominator. That makes further averaging over variables ψ, ϕ, ν in the analytical form practically impossible. Therefore the problem set will be solved for a particular case - for a satellite of right circular cylindrical shape.

Making integration for the surface of a right circular cylinder in formulae (63) we obtain

$$I_1(\delta) = b_0\sin\delta + d_0|\cos\delta| \quad ;$$

$$I_2(\delta) = d_0\sin\delta \;\; \sin(\cos\delta) \quad ;$$

$$I_3(\delta) = a_0\sin\delta + c_0|\cos\delta| + k_0 \quad ;$$

$$I_4(\delta) = 0 \quad ;$$

$$I_5(\delta) = 2a_0\sin\delta + c_0|\cos\delta| + k_0 \quad , \qquad (71)$$

where

$$a_0 = (1-\varepsilon)\left\{\frac{2}{3} R_0^3\ell + \frac{1}{6} R_0\ell^3\right\} + \frac{2}{9}\varepsilon R_0\ell^3 \quad ;$$

$$b_0 = 2(1-\varepsilon)R_0^3\ell \quad ;$$

$$c_0 = (1-\varepsilon)\left\{\frac{\pi R_0^4}{2} + \frac{\pi R_0^2\ell^2}{4}\right\} + \varepsilon\pi R_0^4 \quad ;$$

$$d_0 = (1-\varepsilon)\,\frac{\pi R_0^4}{4} \quad ;$$

$$k_0 = (1-\varepsilon)\,\frac{v}{V}\,\frac{\pi R_0^4}{4} \quad . \qquad (72)$$

Substitute Eq. (71) into (68)

$$(\vec{M}_2)_{\vec{L}} = \rho VL \Big\{ -\sin\delta \left(\frac{1}{A}\, 2a_0 \sin^2\tilde{\theta} + \frac{1}{C}\, b_0 \cos^2\tilde{\theta}\right)$$

$$- |\cos\delta| \left(\frac{1}{A}\, c_0 \sin^2\tilde{\theta} + \frac{1}{C}\, d_0 \cos^2\tilde{\theta}\right)$$

$$+ \sin\delta\cos^2\beta \cdot \frac{1}{A}\, a_0 \sin^2\tilde{\theta}$$

$$+ \sin\delta\sin\cdot(\cos\delta)\cdot\cos\beta\cdot\frac{1}{A}\, d_0 \cos\tilde{\theta}\sin\tilde{\theta}$$

$$- \frac{1}{A}\, k_0 \sin^2\tilde{\theta} \Big\} \quad . \tag{73}$$

According to Ref. (2), in addition to Eq. (59) we take the following approximation

$$\sin\delta\sin(\cos\delta) \simeq 2\sin\delta\cos\delta \quad . \tag{74}$$

In terms of Eqs. (59) (74) and exact Eq. (69) Eq. (73) is transformed into the form

$$(\vec{M}_2)_{\vec{L}} = \rho VL \Big\{ -\frac{1}{A}\,(2a_0+k_0) + \left(\frac{2a_0}{A} + \frac{k_0}{A} - \frac{b_0}{C}\right)\cos^2\tilde{\theta}$$

$$+ \left[\frac{1}{A}\,(2a_0+k_0-c_0) + \left(\frac{1}{C}\,(2a_0+k_0-c_0)\right)\cos^2\delta\right]$$

$$\times \cos^2\delta + \frac{a_0}{A}\,(e_{V_3} - \cos\tilde{\theta}\cos\delta)^2$$

$$+ \frac{2d_0}{A}\,\cos\tilde{\theta}\cos\delta\,(e_{V_3} - \cos\tilde{\theta}\cos\delta) \Big\} \quad , \tag{75}$$

where

$$\cos\tilde{\theta} = (\vec{e}_{L_3}\cdot\vec{k}) = k_3(\phi,\theta) \quad . \tag{75'}$$

Formulae (38), (37), (32), (75') are inserted into (75) and expression (75) is averaged over the variables θ, ϕ, ν

$$\overline{(\vec{M}_2)_{\vec{L}}}^{\psi,\phi,\nu} = \rho_\pi \sqrt{\frac{\mu}{p}}\, L\Bigg\{-\frac{1}{A}(2a_0+k_0)\overline{\bar{\rho}\bar{V}}^{\nu}$$

$$+\left(\frac{2a_0}{A}+\frac{k_0}{A}-\frac{b_0}{C}\right)\overline{\bar{\rho}\bar{V}}^{\nu}\,\overline{\cos^2\tilde{\theta}}^{\phi}+\frac{1}{A}(2a_0+k_0-c_0)$$

$$\times\ \overline{\bar{\rho}\bar{V}\cos^2\delta}^{\psi,\phi,\nu}+\left[\frac{1}{C}(b_0-d_0)-\frac{1}{A}(2a_0+k_0-c_0)\right]$$

$$\times\ \overline{\bar{\rho}\bar{V}\cos^2\tilde{\theta}\cos^2\delta}^{\psi,\phi,\nu}+\frac{a_0}{A}\overline{\bar{\rho}\bar{V}e_{V_3}^2}^{\nu}$$

$$-\frac{2a_0}{A}\overline{\bar{\rho}\bar{V}e_{V_3}\cos\tilde{\theta}\cos\delta}^{\psi,\phi,\nu}+\frac{a_0}{A}\overline{\bar{\rho}\bar{V}\cos^2\tilde{\theta}\cos^2\delta}^{\psi,\phi,\nu}$$

$$+\frac{2d_0}{A}\overline{\bar{\rho}\bar{V}e_{V_3}\cos\tilde{\theta}\cos\delta}^{\psi,\phi,\nu}-\frac{2d_0}{A}\overline{\bar{\rho}\bar{V}\cos^2\tilde{\theta}\cos^2\delta}^{\psi,\phi,\nu}\Bigg\} \tag{76}$$

where

$$\overline{\cos^2\tilde{\theta}}^{\phi} = E_2 \quad ;$$

$$\overline{\bar{\rho}\bar{V}\cos^2\delta}^{\psi,\phi,\nu} = \overline{\bar{\rho}\bar{V}(e_{V_1}^2+e_{V_2}^2)}^{\nu}E_1 + \overline{\bar{\rho}\bar{V}e_{V_3}^2}^{\nu}E_2 \quad ;$$

$$\overline{\bar{\rho}\bar{V}e_{V_3}\cos\tilde{\theta}\cos\delta}^{\psi,\phi,\nu} = \overline{\bar{\rho}\bar{V}e_{V_3}^2}^{\nu}E_2 \quad ;$$

$$\overline{\bar{\rho}\bar{V}\cos^2\tilde{\theta}\cos^2\delta}^{\psi,\phi,\nu} = \overline{\bar{\rho}\bar{V}(e_{V_1}^2+e_{V_2}^2)}^{\nu}H_1 + \overline{\bar{\rho}\bar{V}e_{V_3}^2}^{\nu}H_2 \quad ;$$

$$H_1 = \frac{1}{16}\sin^4\kappa\sin^2\theta(1+3\cos\theta)+\frac{1}{4}\sin^2\kappa\cos^2\kappa\sin^4\theta$$

$$-\sin^2\kappa\cos^2\kappa\cos\theta\sin^2\theta+\cos^2\kappa\cos^2\theta E_1 \quad ;$$

$$H_2 = \frac{3}{8}\sin^4\kappa\sin^4\theta+\frac{1}{2}\cos^2\kappa\sin^2\kappa\cos^2\theta\sin^2\theta$$

$$+2\sin^2\kappa\cos^2\kappa\cos\theta\sin^2\theta+\cos^2\kappa\cos^2\theta E_2 \quad ;$$

$$E_1 = \frac{1}{4}\sin^2\kappa(1+\cos^2\theta) + \frac{1}{2}\cos^2\kappa\sin^2\theta \quad ;$$

$$E_2 = \frac{1}{2}\sin^2\kappa\sin^2\theta + \cos^2\kappa\cos^2\theta \quad ;$$

$$\overline{\bar{\rho}\bar{V}(e^2_{V_1}+e^2_{V_2})}^{\nu} = (m_1^2+m_2^2)K_1 + (\ell_1^2+\ell_2^2)K_2 \quad ;$$

$$\overline{\bar{\rho}\bar{V}e^2_{V_3}}^{\nu} = m_3^2 K_1 + \ell_3^2 K_2 \quad ;$$

$$K_1 = \frac{(1-e^2)^{3/2}}{2\pi}\int_0^{2\pi}\frac{\bar{\rho}(\nu)(e+\cos\nu)^2 d\nu}{\sqrt{1+e^2+2e\cos\nu}\;(1+e\cos\nu)^2} \quad ;$$

$$K_2 = \frac{(1-e^2)^{3/2}}{2\pi}\int_0^{2\pi}\frac{\bar{\rho}(\nu)\sin^2\nu d\nu}{\sqrt{1+e^2+2e\cos\nu}\;(1+e\cos\nu)^2} \quad . \tag{76'}$$

Moment $\vec{M}_3$

The necessity of allowing for the variation in the atmosphere density within the satellite's length (the possibility of the existence of aerogradient moment) was evident long ago, see e.g., (11). In (3) the problem has been considered in detail, whereas here the main results are presented in brief.

Let us resolve expression (12') over the unit vectors of the semi-connected system of coordinates (Fig. 3). Since the moment of aerogradient is a correction of the first order of smallness to the moment of friction some simplifying suggestions can be made: in (12') let $\varepsilon = v/V = 0$ and assume $\vec{e}_V \perp \vec{e}_r$ where $\vec{e}_V$ and $\vec{e}_r$ are unit vectors of the velocity and of geocentric radius vector of the satellite, respectively. Equations (16) and (17) are considered, and $\vec{M}_3$ is projected on to the vector $\vec{L}$, then the following expression for $(\vec{M}_3)_{\vec{L}}$ is derived

$$(\vec{M}_3)_{\vec{L}} = -\frac{\rho V^2}{H}\{\sin\theta\sin\beta\sin\alpha\cdot\phi_1(\delta)$$

$$+ (\cos\delta\sin\theta\cos\beta - \sin\delta\cos\theta)\cos\alpha\cdot\phi_2(\delta) \quad ,(77)$$

where

$$\phi_1(\delta) = -\sin^3\delta\int_{s^*} n_y z^2 ds + \sin^2\delta\cos\delta\int_{s^*}(2n_y yz - n_z z^2)ds$$

$$+ \sin\delta\cos^2\delta\int_{s^*}(2n_z yz - n_y y^2)ds$$

$$- \cos^3\delta\int_{s^*} n_z y^2 ds \quad ,$$

$$\phi_2(\delta) = \sin\delta\int_{s^*} n_y x^2 ds + \cos\delta\int_{s^*} n_z x^2 ds \quad , \quad (78)$$

the angle α is shown in Fig. 7 (the system xyz is semi-connected) and the angle β in Fig. 6.

When the satellite is a right circular cylinder the formulae take the form

$$\phi_1(\delta) = -\left\{\frac{R_0 \ell^3}{6}\sin^3\delta + \frac{\pi R_0^2\ \ell^2}{4}\sin^2\delta|\cos\delta|\right.$$

$$+ \frac{4}{3}R_0^3\ell\sin\delta\cos^2\delta + \frac{\pi R_0^4}{4}\cos^2\delta|\cos\delta| \quad ;$$

$$\phi_2(\delta) = \frac{2}{3}R_0^3\ell\sin\delta + \frac{\pi R_0^4}{4}|\cos\delta| \quad . \quad (79)$$

where ℓ and R_0 are the length and radius of the cylinder.

The averaging of expression (77) over variables ψ, ϕ, ν faces considerable difficulties since the functions $\sin\beta$, $\sin\alpha$, $\cos\alpha$ cannot be presented analytically as polynomials of power not higher than the third in $\cos\delta$. The moment of aerogradient can only be considered exactly by numerical methods. An approximate estimation of the upper boundary of the moment of aerogradient will be given here.

Let a satellite be a right circular cylinder, the relation of the satellite's radius R_0 to the length ℓ being $R_0/\ell \ll 1$ (rocket-carrier). Only the terms of the form $R_0\ell^3$ will be taken into account in Eq. (79), the rest of the terms being neglected because of their higher order of smallness. Then

$$(\vec{M}_3)_{\vec{L}} = \frac{\rho V^2}{H} \sin\theta\sin\beta\sin\alpha \frac{R_0\ell^3}{6} \sin^2\delta \quad . \qquad (80)$$

Tumbling is a stable regime of motion of a dynamically extended satellite around the mass center, whence $\sin\theta \cong 1$. The upper boundary of the moment of aerogradient will be estimated by putting $\sin\beta = 1$, $\sin\alpha = 1$, (this case is consistent with the satellite's tumbling in the orbit plane).

An approximation is made, the error not exceeding 14%

$$\sin^3\delta \simeq \sin^2\delta = 1-\cos^2\delta \quad . \qquad (81)$$

On inserting Eq. (81), (39), (37) into (80) and on making the averaging over the variables ψ, ϕ, ν we obtain

$$\overline{(\vec{M}_3)_{\vec{L}}}^{\psi,\phi,\nu} = \frac{\rho_\pi \mu}{Hp} \cdot \frac{R_0\ell^3}{6} \{G_1 - \overline{\bar{\rho}\bar{V}^2 (e^2_{V_1} + e^2_{V_2})}^{\nu}$$

$$\times E_1 - \overline{\overline{\rho \bar{V}^2 e^2_{V_3}}}^{\,\nu} E_2\}$$

$$= \frac{\rho_\pi \mu}{Hp} \cdot \frac{R_0 \ell^3}{6} \{G_1 - \left[(m_1^2+m_2^2) I_1 + (\ell_1^2+\ell_2^2) I_2\right] E_1$$

$$- \left[m_3^2 I_1 + \ell_3^2 I_2\right] E_2\} \quad , \tag{82}$$

where E_1 and E_2 are expressed by formulae (76'); G_1 is given in terms of Eq. (56) and I_1, I_2 are presented by Eq. (48).

Moment of Magnetic Forces

The moment of forces affecting the rotating satellite in the Earth's magnetic field is considered extensively in Ref. 14. In order to solve our problem we start with the formula for relative drag of an empty cylindrical satellite which is rotating relative to a transverse axis (4)

$$- \frac{1}{\omega} \frac{d\omega}{dt} = \gamma = \frac{0.9}{2C^2} \frac{\sigma R_0^2}{\rho_M \ell^2} (H_{\dagger} \sin\eta)^2 \quad , \tag{83}$$

where γ is relative spin acceleration ($\omega = \omega_0 e^{-\gamma t}$), $H_{\dagger}$ is the Earth's magnetic field intensity (in oersteds), η is an angle between the directions $\vec{H}_{\dagger}$ and $\vec{L}$, $\cos\eta = (\vec{e}_H \cdot \vec{e}_L)$, σ is the electrical conductivity of the material of the satellite's body in the CGS system, $C = 3\times 10^{10}$ cm/sec; ρ_M is the density of this material, R_0 and ℓ are the radius and length respectively of the satellite.

Equation (83) makes no allowance for the satellite's proper rotation defined by

$$\dot{\phi} = \frac{A-C}{C}\,\dot{\psi}\cos\theta \quad , \tag{84}$$

where A and C are principal moments of inertia of the axisymmetric satellite. However in the motion of a long cylindrical satellite the angle θ is practically equal to 90° (see Ref. 6), so that the satellite drag in the magnetic field is described fairly precisely by Eq. (83). According to Ref. (10) the following expression is taken for the vector of the Earth's magnetic field intensity $\vec{H}_{\oplus}$

$$\vec{H}_{\oplus} = \frac{\mu_E}{r_{\oplus}^{\,3}}\,\{\vec{k}_E - 3(\vec{k}_E\cdot\vec{e}_r)\vec{e}_r\} \quad , \tag{85}$$

$\vec{e}_r$ and $r_{\oplus}$ are respectively a unit vector and a modulus of the satellite's geocentric radius vector in a fixed geocentric equatorial system of coordinates (see Fig. 8), $\vec{k}_E$ is a unit vector directed towards the north magnetic pole of the Earth, $\mu_E = 8\times10^{25}$ oersted cm^3 is the value of the magnetic moment of the terrestrial dipole.

The relative drag of the satellite at a point of its orbit is determined by averaging Eq. (83) and taking account of Eq. (85) over the satellite's orbital motion. To this end let the vector $\vec{e}_r$ be presented as a function of the true anomaly ν. Following Fig. 8 the vector $\vec{e}_r$ can be expressed in the form

$$\vec{e}_r = \vec{e}_\pi\cos\nu + \vec{e}_{90}\sin\nu \tag{86}$$

where $\vec{e}_\pi$ is the unit vector of geocentric perigee direction, and $\vec{e}_{90}$ is a unit vector corresponding to the direction $\nu = 90°$; the vectors $\vec{e}_\pi$ and $\vec{e}_{90}$ have the following components in the equatorial system of coordinates

$$\vec{e}_\pi = \Big[(\cos\Omega\cos\omega - \sin\Omega\cos i\sin\omega) \quad ;$$

$$(\sin\Omega\cos\omega + \cos\Omega\cos i\sin\omega); \quad \sin i\sin\omega\Big] \quad ,$$

$$\vec{e}_r = \Big[(-\cos\Omega\sin\omega - \sin\Omega\cos i\cos\omega) \quad ;$$

$$(-\sin\Omega\sin\omega + \cos\Omega\cos i\cos\omega); \quad \sin i\cos\omega\Big] \quad . \tag{87}$$

The vector $\vec{k}_E$ makes an angle of 11° with the direction towards the North pole and completes a full revolution around this direction within a day. As is shown by numerical calculations one can believe with accuracy to 1-2% that $\vec{k}_E$ is directed towards the celestial pole.

The satellite's relative drag γ_{magn} being averaged over the orbital motion can be derived as follows

$$\gamma_{magn} = \frac{(1-e^2)^{3/2}}{2\pi} \int_0^{2\pi} \frac{\gamma(\nu)d\nu}{(1+e\cos\nu)^2}$$

$$= \frac{(1-e)^{3/2}}{2\pi} \frac{9\sigma R_0^3}{2C^2\rho_M\ell^2} \int_0^{2\pi} \frac{H_{\vec{0}}^2(\nu)\sin^2\eta(\nu)d\nu}{(1+e\cos\nu)^2} \quad , \tag{88}$$

where

$$\vec{e}_H = \frac{\vec{H}_{\vec{0}}}{H_{\vec{0}}} \quad ; \quad \sin\eta = \sqrt{1-(\vec{e}_H\vec{e}_L)^2} \quad ;$$

$$\vec{e}_L = (\cos\delta_L\cos\alpha_L \ ; \ \cos\delta_L\sin\alpha_L \ ; \ \sin\delta_L) \quad .$$

Estimation and Comparison of Effects of Moments

Let us carry out an absolute and comparative estimation of the effect of the above moments on the satellite's spin rate.

The derivative with respect to time of the modulus of the satellite's angular moment $|\vec{L}|$ is equal to the projection of the resultant perturbation moment on to the axis $\vec{L}$:

$$\frac{dL}{dt} = (\vec{M})_{\vec{L}} = \sum_{i=1}^{n} (\vec{M}_i)_{\vec{L}} \quad . \tag{89}$$

We express the relative variation in terms of L:

$$\frac{1}{L}\frac{dL}{dt} = \frac{(\vec{M})_{\vec{L}}}{L} = \sum_{i=1}^{n} \frac{(\vec{M}_i)_{\vec{L}}}{L} \quad . \tag{90}$$

Consider the motion of the satellite over a time interval within which the variation of L is small compared with the value of L:

$$\frac{\Delta L}{L} << 1 \quad . \tag{91}$$

In this case it follows from Eq. (90) that

$$L = L_0 \exp \frac{(\vec{M})_{\vec{L}}}{L} t = L_0 \exp \sum_{i=1}^{n} \frac{(\vec{M}_i)_{\vec{L}}}{L} t \quad ,$$

or on introducing the designations

$$\frac{(\vec{M}_i)_{\vec{L}}}{L} = -\gamma_i \quad ; \quad \frac{(\vec{M})_{\vec{L}}}{L} = \gamma_{real} \quad , \tag{92}$$

$$L = L_0 \exp(-\gamma_{real} \cdot t) = L_0 \exp\left[-\sum_{i=1}^{n} \gamma_i t\right] \quad , \tag{93}$$

where γ_{real} is a drag value really observed. It follows from Eq. (93)

$$\gamma_{real} = \sum_{i=1}^{n} \gamma_i \quad .$$

If one takes into account that (see Ref. 1) $L = \dot{\psi}A$ (where $\dot{\psi}$ is the satellite's spin rate with respect to the vector $\vec{L}$, A is the moment of inertia relative to the satellite's transverse axis), this (93) implies that the variation of spin-rate of the satellite actually observed is a result of the total effect of all the perturbing moments whereas γ is an additive value. γ_{real} can be obtained from observations, in particular photometric observations. γ^{-1} has a time-like dimension and it can be called the characteristic time of the satellite drag as affected by some relevant moment.

We shall estimate the moments. For a particular example let us take a rocket-carrier of the Third Soviet Artificial Earth Satellite ($1958\delta_1$), the spin of which has been studied rather extensively.

Choose the rocket parameters as follows: diameter $2R_0$ and length ℓ corresponding to the rocket-carrier of the space-vehicle "Vostok" i.e. 2.5m and 28m respectively (7), mass cross-section relation in accordance with the definition by Priester and Martin (8) i.e. equal to 125 kg/m^2, mean cross-sectional area derived from King-Hele's formula (9)

$$\frac{2}{\pi}(2\ell R_0 + \pi R_0^2) \quad . \tag{94}$$

The values of 0.02 and 1000 m/sec will be taken for the parameters ε and v respectively (2).

In accordance with the data of Ref. (6) we adhere to the point that the satellite's rotation is a mere tumbling in which case $\cos\theta = 0$, $\sin\theta = 1$. Let the satellite's spin rate be equal to 0.35 rad/sec. Insofar as the AES $1958\delta_1$ was a fairly extended and uniform cylinder, the distance $R_{Y'}$ contained between the axes of geometric and dynamic symmetry and the angle κ made between those axes seemed to be small values. Assume (Eq. 95)

$$R_{Y'} = 0.1\text{m} \quad , \quad \sin\kappa = 0.1 \quad .$$

Note that the approximate equations below hold true for a satellite in an elliptic orbit with eccentricity of about 0.1 (this is the orbit eccentricity of satellite $1958\delta_1$);

$$I_1 \simeq 0.1 \;\; ; \;\; J_1 \simeq 0.1 \;\; ; \;\; K_1 \simeq 0.1 \;\; ; \;\; G_1 \simeq 0.1 \;\; ;$$

$$I_2 \simeq 0 \;\; ; \;\; J_2 \simeq 0 \;\; ; \;\; K_2 \simeq 0 \;\; ; \;\; G_2 \simeq 0.1 \qquad (96)$$

(see Eq. (48) for I_1, I_2, J_1, J_2, (76') for K_1, K_2 and Eq. (43) for $\bar{\rho}(\nu)$).

Estimate the contribution of the moment of completely inelastic auto-rotation $\vec{M}_{11}$ to the variation of the value of the satellite's angular moment.

Inserting the above values R_0 and ℓ into Eq. (60) we find that for $1958\delta_1$

$$F_{V_2} \simeq -70\text{m}^2 \quad . \qquad (97)$$

Substituting Eq. (46) for numerical values from Eqs. (94-97) and using Eqs. (47), (31), it can be found for $1958\delta_1$ that

$$\overline{(\vec{M}_{11})_{\vec{L}}}^{\psi,\phi,\nu} = 2.24\times10^{-4} \sin^2\theta\cos\theta \;\text{kg}\cdot\text{m}^2/\text{sec}^2 \quad .$$

On the assumptions given the moment of inertia of the satellite with respect to the transverse axis $A = 3.9\times 10^5$ kg·m² and the angular moment

$$L = 1.36\times10^5 \text{ kg}\cdot\text{m}^2/\text{sec} \quad . \tag{98}$$

In such a case for the moment of completely inelastic auto-rotation

$$\gamma_{11} = \sin^2\Theta\cos\Theta\cdot1.64\times10^{-9}\ \text{sec}^{-1} \quad ;$$

$$\gamma_{11}^{-1} = -\frac{1}{\sin^2\Theta\cos\Theta}\cdot7.1\times10^3 \text{ days} \quad . \tag{99}$$

Similarly estimating the contribution of the diffuse-mirror auto-rotation moment $\vec{M}_{12}$ we obtain

$$\gamma_{12} = -\cos\Theta\cdot1.54\times10^{-10}\ \text{sec}^{-1} \quad ,$$

$$\gamma_{12}^{-1} = (-\cos\Theta)^{-1}\cdot7.5\times10^4 \text{ days} \quad . \tag{100}$$

As is clearly illustrated by Eqs. (99) and (100), the satellite's rotation can be either decelerated or accelerated by auto-rotation moments ($\vec{M}_{11}$ and $\vec{M}_{12}$, Eqs. (46) and (55)) depending on the sign of $\cos\Theta$. The effect of the moment $\vec{M}_2$ of friction being estimated, we can obtain

$$\gamma_2 = 9.4\times10^{-10}\ \text{sec}^{-1} \ ; \quad \gamma_2^{-1} = 1.3\times10^4 \text{ days} \ . \tag{101}$$

For the moment of aerogradient $\vec{M}_3$

$$\gamma_3 = -1.3\times10^{-10}\ \text{sec}^{-1} \qquad \gamma_3^{1} = -8.9\times10^4 \text{ days} \ . \tag{101'}$$

It may be noted that the moment $\vec{M}_3$ (Eq. 77) depending on the signs of trigonometric functions of the angles α and β as well as $\vec{M}_{11}$, $\vec{M}_{12}$ can either drag or accelerate the satellite's rotation. Supposing the satellite of the above parameters is tumbling in the orbit plane, the numerical estimations of the aerogradient effect will show (3) that when the satellite's period is equal to 20 sec, the relation $M_3:M_2 \approx 0.2$ and it grows with increase in the rotational period.

It must be borne in mind that for the 1957β satellite the precession period was of the order of 200 sec. However, it is to be emphasized once again that these estimations of the moment $\vec{M}_3$ are maximum values.

The values γ_{magn} and γ^{-1}_{magn} for the magnetic moment depend significantly on the characteristics of the material of which the satellite is made. They depend linearly on the electrical conductivity σ of the satellite's body and on its specific volume $1/\rho_M$. Depending on the material, the value $1/\rho_M$ may vary by 3-4 times, whereas the value σ may range widely. The results of numerical calculations of γ^{-}_{magn} values for the 1958δ_1 satellite are summarized in Table 1 (in cases of aluminum and alloyed steel at different orientations of the satellite's axis of roation).

Table 1.

Revolution No.	Satellite's Material	$\alpha_L=0^0$ $\delta_L=0^0$	$\alpha_L=90^0$ $\delta_L=0^0$	$\alpha_L=90^0$ $\delta_L=90^0$
1074	Steel	5299 days	3936 days	4856 days
	Aluminum	124	92	114
1786	Steel	4945	3655	4515
	Aluminum	115	85	105

As is shown by the above estimations of aerodynamic moments for the 1958δ_1 satellite, the values γ_i corresponding to the moments of completely inelastic auto-rotation, friction, diffuse-mirror auto-rotation and aerogradient are related to each other as $12.6\sin^2\Theta\cos\Theta$; $7.2 : 1.2\cos\Theta : 1.0$. The values of γ_i corresponding to the auto-rotation moments depend on the factor of $R_{y'}\sin\kappa$ linearly, so that when the axes of the satellite's geometric and dynamic symmetry are disposed very close to each other or even coincide, the effect of the auto-rotation moment becomes quite small or vanishes altogether.

As is shown in Table 1, γ_{magn} and γ^{-1}_{magn} characterizing magnetic drag depend strongly on the material of the satellite but vary slightly over long time intervals; thus the satellite's magnetic drag can be considered practically constant over long time intervals.

It can also be inferred from Table 1 that within the framework of the above reasoning γ^{-1}_{magn} for aluminum is not consistent with the 1958δ_1 satellite drag actually observed (6) (γ^{-1}_{real} is many times greater than γ^{-1}_{magn} for aluminum).

On Atmosphere Density Calculations From Satellite's Spin Drag

Write down Eq. (94) in the expanded form

$$\gamma_{real} = \gamma_{11}+\gamma_{12}+\gamma_2+\gamma_3+\gamma_{magn} \tag{102}$$

where γ_{11} and γ_{12} correspond to the moments of completely inelastic and diffuse-mirror auto-rotation, and γ_2, γ_3, γ_{magn} to the moments of friction, aerogradient and the magnetic moment respectively. Inserting Eq. (89) into (92) and considering that $L = d\psi/dt \cdot A$, we obtain

$$\gamma_{real} = \frac{d^2\psi/dt^2}{d\psi/dt} \quad . \tag{103}$$

The angular velocity of the satellite's spin $d\psi/dt$ is connected with the satellite's period of rotation P by the relationship $d\psi/dt = 2\pi/P$.

Thus

$$\gamma_{real} = \frac{1}{P}\frac{dP}{dt} \quad . \tag{104}$$

The value of γ_{real} is expressed by Eq. (104) in terms of the values observed, P and dP/dt.

On the other hand

$$\gamma_{11} = \frac{1}{L}\,\overline{(\vec{M}_{11})_{\vec{L}}}^{\psi,\phi,\nu} \quad ; \quad \gamma_{12} = \frac{1}{L}\,\overline{(\vec{M}_{12})_{\vec{L}}}^{\psi,\phi,\nu} \quad ;$$

$$\gamma_{2} = \frac{1}{L}\,\overline{(\vec{M}_{2})_{\vec{L}}}^{\psi,\phi,\nu} \quad ; \quad \gamma_{3} = \frac{1}{L}\,\overline{(\vec{M}_{3})_{\vec{L}}}^{\psi,\phi,\nu} \quad ;$$

$$\gamma_{magn} \simeq \text{const.} \quad .$$

The values of γ_i can be presented in the form

$$\gamma_{11} = \frac{1}{L}\,\rho_\pi\gamma'_{11} \quad ; \quad \gamma_{12} = \frac{1}{L}\,\rho_\pi\gamma'_{12} \quad ;$$

$$\gamma_{2} = \rho_\pi\gamma'_{2} \quad ; \quad \gamma_{3} = \frac{1}{L}\,\rho_\pi\gamma'_{3} \quad ; \tag{105}$$

$$\gamma_{magn} \simeq \text{const.}$$

where

$$\gamma'_{11} = \frac{1}{\rho_\pi} \overline{(\vec{M}_{11})_{\vec{L}}}^{\psi,\phi,\nu} \quad ; \quad \gamma'_{12} = \frac{1}{\rho_\pi} \overline{(\vec{M}_{12})_{\vec{L}}}^{\psi,\phi,\nu} \quad ;$$

$$\gamma'_{2} = \frac{1}{\rho_\pi L} \overline{(\vec{M}_{2})_{\vec{L}}}^{\psi,\phi,\nu} \quad ; \quad \gamma'_{3} = \frac{1}{\rho_\pi} \overline{(\vec{M}_{3})_{\vec{L}}}^{\psi,\phi,\nu} \quad ;$$

γ'_{11}, γ'_{12}, γ'_{2}, γ'_{3} and γ_{magn} are not dependent on ρ_π (see Eqs. (46, 55, 76, 80).

Inserting Eq. (105) into (102) and solving the equation derived with respect to ρ_π yields

$$\rho_\pi = (\gamma_{real} - \gamma_{magn}) \left[\frac{1}{L}\left(\gamma'_{11} + \gamma'_{12} + \gamma'_{3}\right) + \gamma'_{2}\right]^{-1} \qquad ,(107)$$

where γ_{real} is determined by Eq. (104), the values of γ'_i being expressed in terms of Eq. (106), and γ_{magn} calculated numerically from Eq. (88).

If the axes of the satellite's geometric and dynamic symmetry coincide

$$\gamma'_{11} = \gamma'_{12} = 0 \quad . \qquad (108)$$

As has been shown above γ_3 is an order less than the value of γ_2, and therefore γ'_3/L is also an order less than γ'_2.

In neglecting the small value of γ'_3/L and considering Eq. (108) the following formula for ρ_π, the density at perigee can be derived from Eq. (107)

$$\rho_\pi = \frac{\gamma_{real} - \gamma_{magn}}{\gamma'_2} \qquad (109)$$

where γ_2' is calculated as $1/\rho_\pi \overline{(\vec{M}_2)_{\vec{L}}}^{\psi,\phi,\nu}$, and $\overline{(\vec{M}_2)_{\vec{L}}}^{\psi,\phi,\nu}$ is expressed in terms of formula (76) with κ is now equal to zero.

The formula (76) as well as the approximate formulae above for the remaining perturbing moments make it possible to make a qualitative assessment of the effect of the corresponding moment on the rotation of the satellite. However, Eq. (76) has been derived in terms of approximations (59) and (74). Therefore the exact values of $\overline{(\vec{M}_2)_{\vec{L}}}^{\psi,\phi,\nu}$ may be determined by Eq. (73) wherein $\tilde{\theta} = \theta$ now, since $\kappa = 0$.

In presenting the angles δ and β through Eqs. (38) and (69) as the functions of ψ, ν angles, the values of $\overline{(\vec{M}_2)_{\vec{L}}}^{\psi,\nu}$ and γ_2' can be calculated by numerical averaging of formula (73) over independent variables ψ, ν (since as $\kappa = 0$, dependence on the angle κ vanishes).

Values of the atmospheric density calculated by formula (109) in averaging $(\vec{M}_2)_{\vec{L}}$ in terms of numerical integration of formula (73) are given in the second column of Table 2 according to Ref. (6) (the values of ρ_0). These densities ρ_0 are calculated with respect to the spin drag of satellite 1958δ obtained from photometric observations (6), reduced to an altitude of 210 km. Since the specific electrical conductivity of the material of the satellite, σ, is unknown, the calculations have been made putting $\gamma_{magn} = 0$. Naturally, it has led to overestimated values of atmospheric density. Therefore further calculations have been made taking into account the value of γ_{magn} in accordance with the data in Table 1 for

Table 2.

Date	$10^{13}\rho_0$ g/cm^3	$10^{13}\rho$ g/cm^3
1958 VII.29,4	4.17	1.90
30,4	5.99	3.70
31,4	9.09	6.75
1958 VIII. 1,4	8.36	6.02
2,4	6.85	4.53
3,4	4.82	2.53
4,4	6.62	4.33
5,4	6.17	3.85

alloyed steel. In other words the full formula (109) has been used for calculating the values of ρ given in the third column of Table 2.

Taking magnetic drag into account the values of ρ_0 have markedly approached those of the density obtained from the satellite's orbital drag (6,8). The fact that ρ_0 in the average is still somewhat greater than real density values easily accounts for the use of the alloyed steel σ-parameter. The values of ρ for the alloy with higher specific electrical conductivity will be less than those presented in Table 2. As to the general run of the ρ values from Table 2, it is in good agreement with solar activity variations within the period considered (see Fig. 9 of (6)).

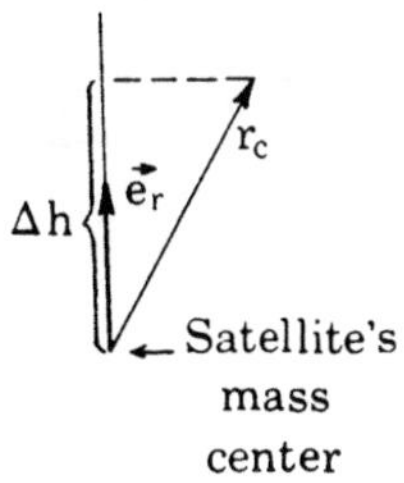

Figure 1.

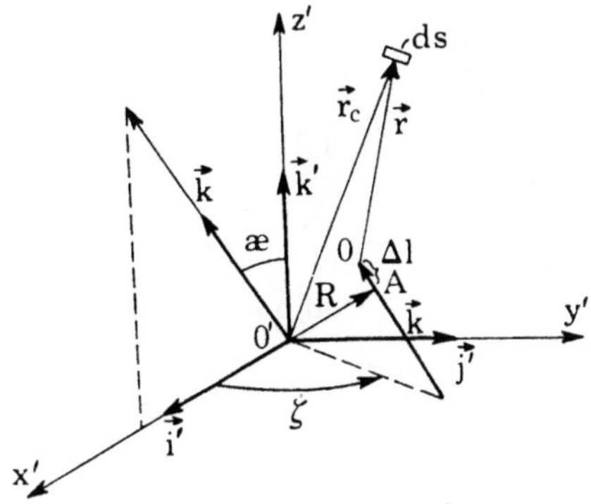

Figure 2.

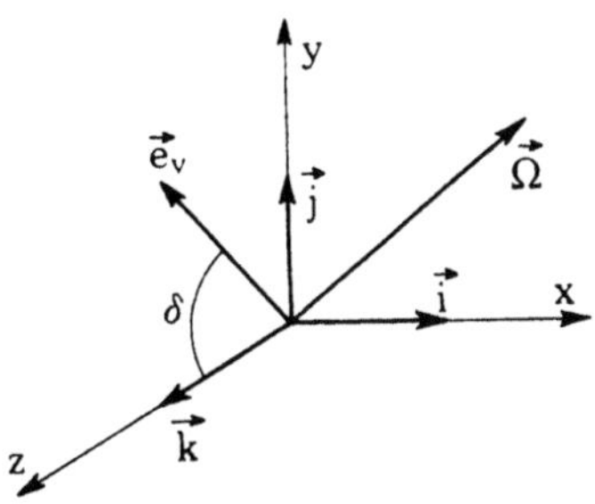

Figure 3.

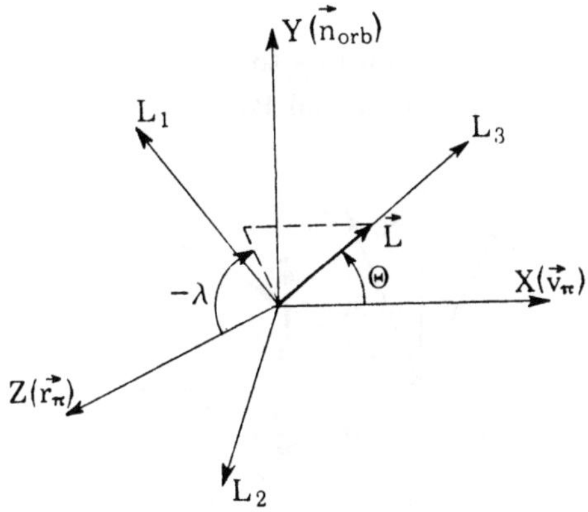

Figure 4.

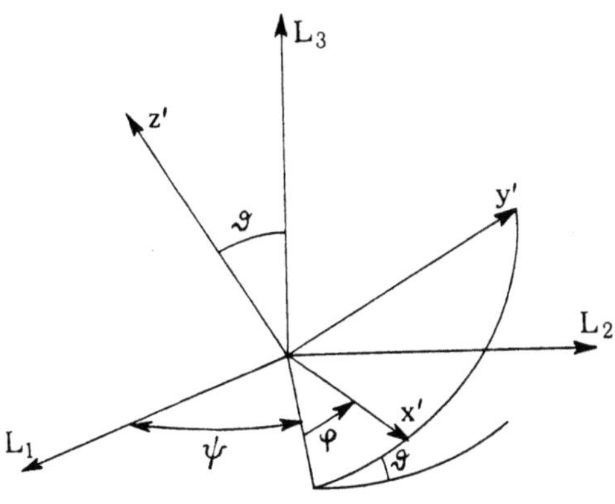

Figure 5.

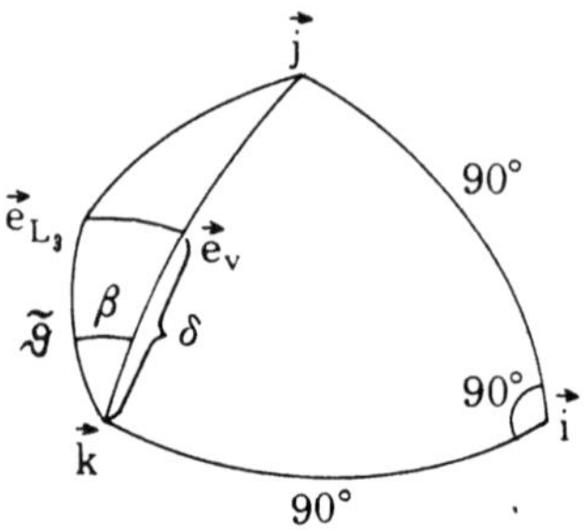

Figure 6.

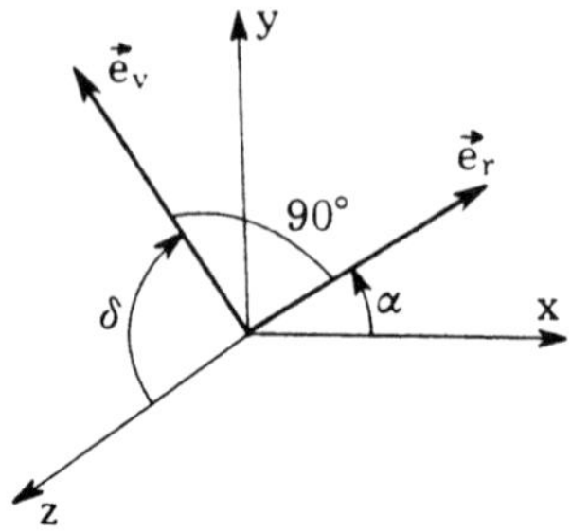

Figure 7.

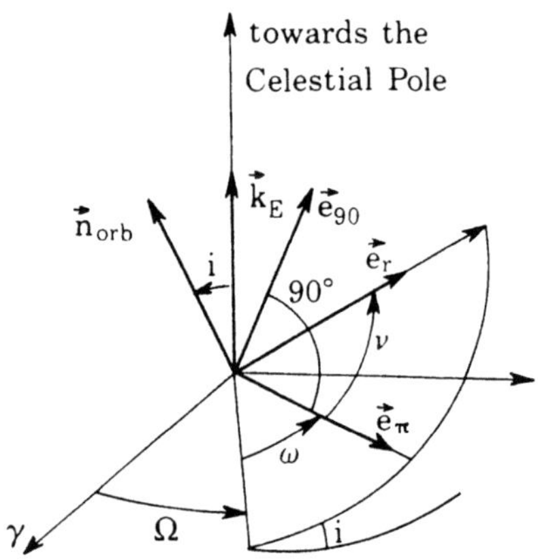

Figure 8.

References

1 Beletskii, V.V.: The motion of a satellite with respect to its center of mass. "Nauka", Moscow, 1965.

2 Beletskii, V.V.; Golubkov, V.V.; Stepanova, E.A.; Hankevitch, I.G.: Determination of the orientation of a satellite from measurement data. IPM AN SSR, Moscow (1968).

3 Kolesnik, S.Ya.; Grigorevsky, V.M.; Beletskii, V.V.: The effect of the aerogradient moment on the motion of a satellite with respect to its center of mass. Buletin stantsii nabludenia ISZ. AN SSSR, Moscow, (1974).

4 Grigorevsky, V.M.; Schmelev, G.M.: Nabludenia ISZ. Vlan-Bator, n. 10 (1970) 257.

5 Beletskii, V.V.: Iskustvenie Sputniki Zemli. AN SSSR, 1st ed., Moscow (1958) 25.

6 Grigorevsky, V.M.; Beletskii, V.V.; Kolesnik, S.Ya.: Nabludenia ISZ. Interkosmos, Praga, n. 12 (1973) 228.

7 Mazokina, G.P.: Astronautiks. Voenizdat, Moscow (1968). (see also: Mond Landing. Dokumentation der Weltraumfahrt, USA and UdSSR. Belser Verlag: Stuttgart, 1970).

8 Priester, W.; Martin, H.A.: Mitteilungen der Universitäts-Sternwarte, Bonn, n. 29 (1960).

9 King-Hele, D.: Orbits of satellites immersed in an atmosphere. Mir, Moscow (1968).

10 Landau, L.D.; Lifchitz, E.M.: Teoria Polia. Fismatgiz: Moscow, 1960.

11 Notny, P.; Oleak, H.: Veröffentlichungen der Sternwarte Babelsberg, Bd. 13, n. 3 (1959).

12 Beletskii, V.V.: The effect of aerodynamic forces on the motion of a satellite with respect to its center of mass. (lecture notes). Moscow State University, 1968-1971.

13 Beletskii, V.V.: Kosmitcheskie Issledovania U.Sh., 2 (1960) 206.

14 Golubkov, V.V.: IPM AN SSSR, Moscow, preprint n. 50 (1971).

Satellite Attitude Control with Desaturated Rotors

H. I. WEBER
Departmento de Engenharia Mecânica
Universidade Estadual de Campinas

Abstract

An active attitude control system with 3 rotors, for positioning an Earth pointing satellite on a circular orbit is presented. The system is subjected only to the influence of the gravitational moment, and impact disturbances or sinusoidal moment perturbations are considered. A different approach is developed for the control. The desaturation of the rotors is done simultaneously with the control of the whole system. The control law is obtained by means of a suitable optimal linear controller and the result is optimal in the sense of a minimum angle deviation from the equilibrium position.

Introduction

Earth pointing satellites on a stable equilibrium position may have the attitude controlled by desaturated rotors. This name denotes rotors that cannot saturate because simultaneously to the control action their angular velocity is maintained in the neighborhood of an equilibrium position. As force field to allow this action, the gravitational moment is used (1). We investigate the problem of a stable satellite on a circular orbit with three rotors, whose axes are parallel to the principal axes of the satellite. The rotors have a zero relative velocity equilibrium position (slow rotors) and are commanded by a linear optimal controller. Two kinds of perturbations will be considered: the initial condition disturbance, due to a meteorite impact (2), and the sinusoidal moment disturbance.

Mathematical Model

To define the equations of motion a body-fixed coordinate system will be used. In the equilibrium position the x-axis is directed along the trajectory (roll axis), the y-axis is directed along the position vector (yaw axis), the z-axis is orthogonal to the orbit plane (pitch axis). The corresponding angles are ϕ, ψ and Θ respectively. Besides Ω represents the angular velocity on orbit, ω the angular velocity of the satellite and ν the relative angular velocity of the rotors. The mathematical model relates all the angles and their derivatives, except the angular position of the rotors, which are cyclic coordinates. Therefore one obtains a 9th order model that may be linearized considering that due to the control the deviations from the equilibrium position will be of a small amount. In this linearized model the pitch motion of 3rd order is uncoupled from the roll-yaw motion of 6th order.

To reduce the number of coordinates where the control actuates, it is convenient to use the following state variables:

$$x_1 = \left[\Theta \; ; \; \frac{\omega_z}{\Omega} + \frac{J_z}{A_z}\frac{\nu_z}{\Omega} \; ; \; \frac{\nu_z}{\Omega}\right]^T$$

$$x_2 = \left[\phi \; ; \; \psi \; ; \; \frac{\omega_x}{\Omega} + \frac{J_x}{A_x}\frac{\nu_x}{\Omega} \; ; \; \frac{\omega_y}{\Omega} + \frac{J_y}{A_y}\frac{\nu_y}{\Omega} \; ; \; \frac{\nu_x}{\Omega} \; ; \; \frac{\nu_y}{\Omega}\right]^T \qquad (1)$$

where A_x is the principal moment of inertia of the satellite about the x-axis and J_x is the moment of inertia of the roll-rotor about its axis.

The complete mathematical model using the adimensional time $\tau = \Omega t$ is

$$x' = A x + B u \tag{2}$$

where

$$A_1 = \begin{bmatrix} 0 & 1 & -a_{5z} \\ -3a_{1z} & 0 & 0 \\ 3a_{1z} & 0 & 0 \end{bmatrix} \qquad B_1 = \begin{bmatrix} 0 \\ 0 \\ 1 \end{bmatrix}$$

$$A_2 = \begin{bmatrix} 0 & -1 & 1 & 0 & -a_{5x} & 0 \\ 1 & 0 & 0 & 1 & 0 & -a_{5y} \\ 3a_{1x} & 0 & 0 & -a_{1x} & 0 & -a_{3y} \\ 0 & 0 & -a_{1y} & 0 & a_{3x} & 0 \\ -3a_{2x} & 0 & 0 & a_{2x} & 0 & a_{4y} \\ 0 & 0 & a_{2y} & 0 & -a_{4x} & 0 \end{bmatrix} B_2 = \begin{bmatrix} 0 & 0 \\ 0 & 0 \\ 0 & 0 \\ 0 & 0 \\ 1 & 0 \\ 0 & \beta \end{bmatrix}$$

where

$$a_{1x} = -\frac{A_z - A_y}{A_x}, \quad a_{2x} = -\frac{A_z - A_y}{A_x - J_x}; \quad a_{5x} = \frac{J_x}{A_x}$$

(x,y,z permuting)

and

$$a_{3x} = J_x \frac{A_z}{A_x A_y} \qquad a_{3y} = J_y \frac{A_z}{A_x A_y}$$

$$a_{4x} = J_x \frac{A_z}{A_x (A_y - J_y)} \qquad a_{4y} = J_y \frac{A_z}{(A_x - J_x) A_y}$$

$$\beta = \frac{J_x (A_x - J_x)}{A_x} \cdot \frac{A_y}{J_y (A_y - J_y)} \quad .$$

The numerical values that were used for the computations are

$$\frac{A_z}{A_x} = \frac{7}{6}, \quad \frac{A_x}{A_y} = 3, \quad J = J_x = J_y = J_z, \quad \frac{J}{A_y} = \frac{1}{10} \; .$$

Control Law

A linear control law is used to define the behavior of the rotors:

$$u = L\,x \quad , \tag{3}$$

where the feedback matrix is determined considering that the solution of the problem (2) should minimize

$$\bar{J} = \int_0^\infty (x^T Q\, x + u^T R\, u)\, dt \quad . \tag{4}$$

This is obtained when

$$L = - R^{-1} B^T K \tag{5}$$

where

$$A^T K + K A + Q - K B R^{-1} B^T K = 0 \quad .$$

The difficulty remains in selecting convenient weighting matrices Q, R. The usual methods were not satisfying, therefore a new linear optimization concept was established. It is based on searching for the Q, R that result in a solution of (2) which minimizes another functional. In this work the optimal linear controller (3) which results in the minimum angle deviation of the satellite

$$J_1 = \max|\Theta| \quad , \quad J_2 = \max(|\phi|,|\psi|) \tag{6}$$

is presented. First is studied the problem (2) with an initial condition in the angular velocity of the satellite, as it happens by the impact of a meteorite. Then the same controller is investigated for the problem

$$x' = A\,x + B(-R^{-1} B^T K x) + C\,w \tag{7}$$

where the perturbation is originated on an external moment that is supposed sinusoidal

$$C_1 = \left[0 \; ; \; (1-a_{5z}) \; ; \; -1\right] \bar{M}_z \cos \omega_\varepsilon \tau/\Omega$$

$$C_{2x} = \left[0 \; ; \; 0 \; ; \; (1-a_{5x}) \; ; \; 0 \; ; \; -1 \; ; \; 0\right] \bar{M}_x \cos \omega_\varepsilon \tau/\Omega \; . \tag{8}$$

The frequency responses for the angular deviations are obtained, and it is shown that the same Q, R result in a, at least, local minimum. In that way, both the homogeneous and the sinusoidal disturbed motions are optimized (3).

Optimal Results

To satisfy the condition (6) it is necessary to establish a restriction on the control moment. It is arbitrarily assumed that

$$\max |u| < u_{limit} = 50 \quad . \tag{9}$$

Furthermore the functional (4) has the form defined by physical reasoning and equivalence considerations (4).

$$\bar{J}_1 = \int_0^\infty (q_1 \Theta^2 + q_2 (\nu_z/\Omega)^2 + u_z^2) dt$$

$$\bar{J}_2 = \int_0^\infty \{q_1 (\phi^2 + \psi^2) + q_2 \left[(\nu_x/\Omega)^2 + (\nu_y/\Omega)^2\right] + u_x^2 + u_y^2\} dt$$

The optimal result for the pitch problem is for q_1 = 8634, q_2 = 0.344. The corresponding pitch angle variation for a unit angular velocity input is shown in Fig. 1 and represents a 53% reduction in the maximum compared with the uncontrolled system. The same control law gives the frequency response for the pitch angle shown in Fig. 2. The overall improvement in the frequency response is obtained for instance with q_1 = 9400, q_2 = 0.001 but it results in a worse behavior when considering the initial condition problem. The improvement of this behavior (q_1 = 8200, q_2 = 1) on the other hand disagrees with (9) and results in a worse frequency response.

The optimal result for the roll-yaw problem is for q_1 = 14510, q_2 = 0. The corresponding angular variations for a unit angular velocity disturbance in the roll axis is shown in Fig. 3 and represents a 44% reduction in the maximum roll angle compared with the uncontrolled system. Figure 4 shows the same for a yaw input and the reduction of the yaw angle is of 79%. In Fig. 5 the frequency response of the roll angle for a roll moment disturbance is represented.

Conclusions

In this work it is shown how to introduce the rotor angular velocity as a control variable and what is the best linear controller when a minimum deviation from the equilibrium position is wanted. A simultaneous optimization of initial condition and frequency response is presented.

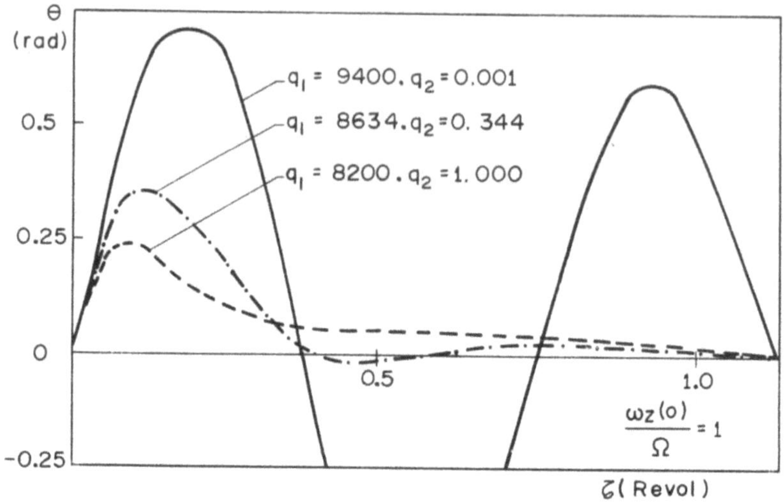

Figure 1. Pitch angle with initial condition disturbance.

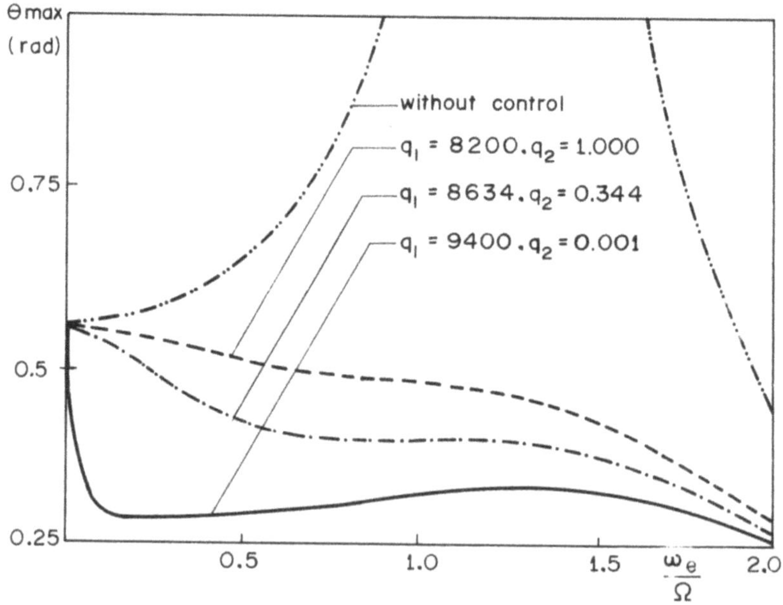

Figure 2. Frequency response for the pitch angle.

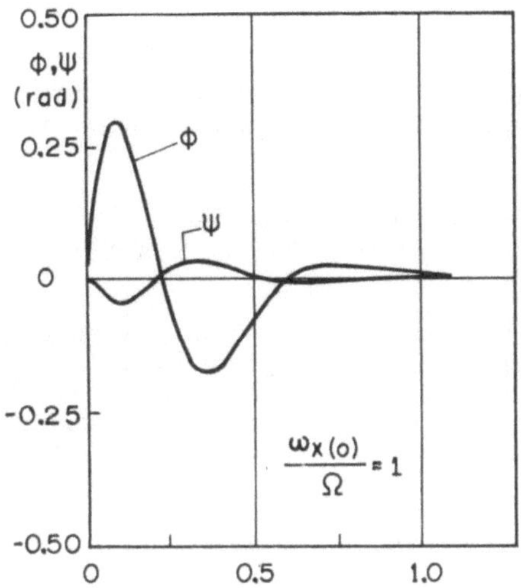

Figure 3. Roll and yaw angles, disturbance on roll-axis.

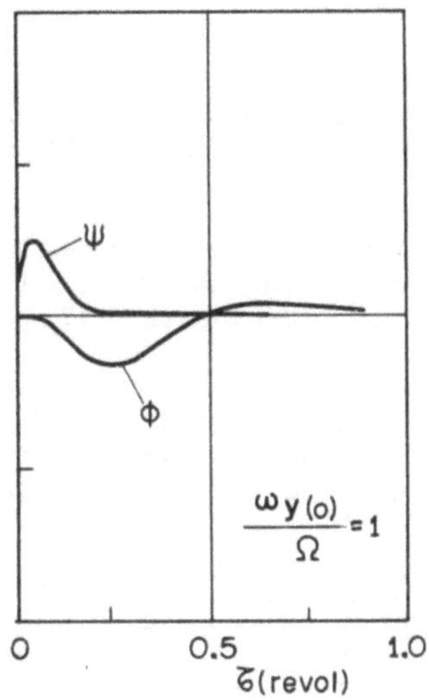

Figure 4. Roll and yaw angles, disturbance on yaw axis.

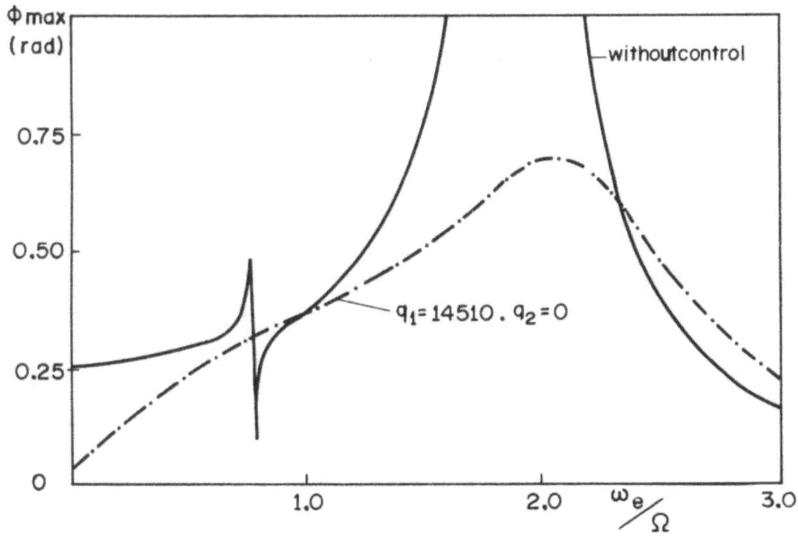

Figure 5. Roll angle frequency response for roll-axis perturbation.

References

1 Weber, H.I.; Shiehlen, W.: Attitude Controllability of a Satellite with Flywheels, J. Spacecraft Rockets 7 n. 4 (1970).

2 Weber, H.I.: Optimale Lageregelung von Satelliten mit Simultan Entdrallten Rotoren. Dr-Ing. thesis, T.U. München, 1971.

3 Weber, H.I.: Otimização Simultânea do Comportamento Dinâmico Próprio e Perturbado, no Projeto de um Sistema Mecânico Linear Ativo, I Simpósio Brasileiro de Ciências Mecânicas, Rio de Janeiro, November 1973.

4 Kreindler, E.; Hedrick, J.K.: On Equivalence of Quadratic Loss Functionals. Int. J. Control 11 n. 2 (1970).

On the Stability of Spinning Flexible Satellites

P. LOHMEIER
Institut B für Mechanik
Tech. Universität München, BRD

Abstract

The attitude stability of spinning flexible satellites in a force-free environment is analyzed. The satellite is modeled by a rigid core body and two symmetric flexible radial beams with rigid tip bodies. Liapunov and Krasovskii stability theorems are employed using the Hamiltonian of the systems as a Liapunov functional. The definiteness checking of the Hamiltonian is complicated due to the fact that it contains the elastic deformations. This problem is solved by applying the modal analysis. Both, symmetric and asymmetric modes are taken into consideration. To determine the eigenfunctions of the rotating beam particular attention has to be paid to the dynamic boundary condition which depends on natural frequencies due to the tip body. The sufficient conditions for asymptotic stability obtained include the well known rigid body stability criteria, the well known "energy-sink" criteria, and in addition, new criteria depending essentially on the flexible properties of the system. Moreover, the influence of the number of the considered eigenfunctions is investigated with respect to the stability boundary.

Introduction

In satellite dynamics the rigid body model has been a useful idealization in numerous investigations during the last decade. The assumption of rigid bodies is questionable, however, with increasing dimensions of the spacecraft. In the case of large flexible beams, serving for example as antennas or part of a scientific equipment, consideration of flexibility is required. To explain the tumbling motion of Explorer I satellite, Thomson and Reiter (1) and Meirovitch (2) have investigated the effect of energy dissipation resulting from the vibration of certain elastic parts of the satellite. On the basis of energy considerations, these investigations conclude that for spin stabilization the spin axis has to be the axis of maximum moment of inertia. Papers by Pringle (3) and Likins (4) establish the usefulness of the Liapunov direct method for the investigation of the attitude stability of satellites, at least for the case of rigid satellites. The Liapunov direct method has been widely used to analyze the stability of discrete systems. In recent years, however, the Liapunov method has been extended to distributed parameter systems. In this connection the investigations of Wang (5) and Parks (6) are mentioned. From these results it can be concluded that one of the major problems in applying the Liapunov direct method to continuous systems is that of constructing a suitable Liapunov functional. Meirovitch and Calico (7) and Likins and Barbera (8) have shown that the Hamiltonian can be used as a Liapunov functional for satellites in a force-free environment.

The purpose of this paper is to investigate the attitude stability of spinning satellites, Fig. 1. The chosen model consists of one rigid core body K_0 spinning around the Z-axis with constant angular velocity Ω. Two flexible beams K_I, K_{II} are attached to the core body. These flexibile beams are orthogonal to the spin axis in the center of mass plane of the undeformed satellite. The deformation in direction of the beam axis is neglected. The essential difference between the models known in literature and the model under consideration are the additional tip bodies K_1, K_2 attached to the beams.

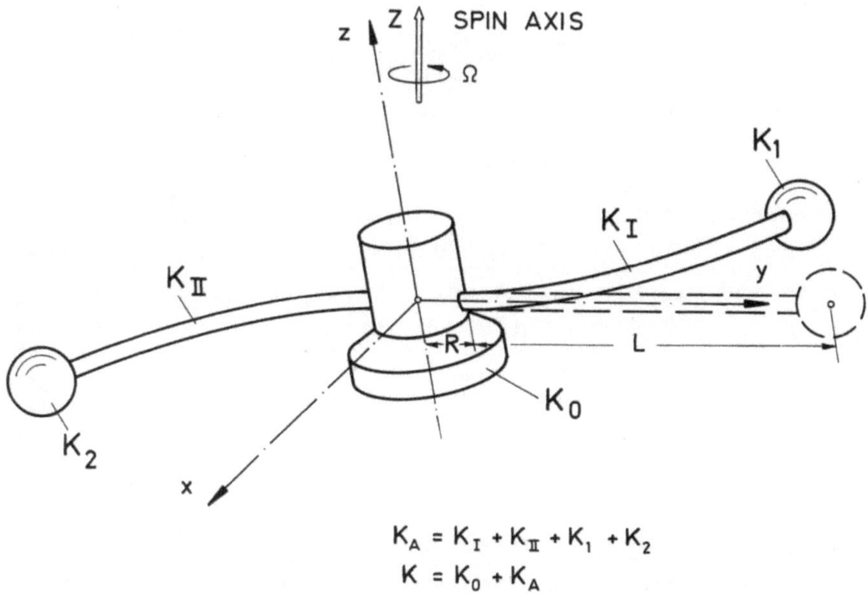

Figure 1. The elastic satellite

For the stability analysis the direct method of Liapunov is used. Assuming energy dissipation as a result of the elastic deformation and considering the force-free system, it can be shown that the Hamiltonian constitutes an appropriate Liapunov functional. Due to the fact that the elastic deformations of the Hamiltonian occur in the integrand it is impossible to determine the definiteness immediately. In this paper the problem can be solved by applying the modal analysis in connection with series expansion. However, the eigenfunctions and eigenfrequencies of the rotating beam depend on the tip body dynamic boundary conditions. Therefore, the usual definitions of self-adjointment, of orthogonality and of the Rayleigh quotient, have to be extended. This leads to sufficient conditions for asymptotic stability in form of infinite series. Detailed numerical results are discussed in the Magnus shape triangle, describing the overall inertia properties of the satellite. Moreover, the influence of the number of the considered eigenfunctions is investigated with respect to the stability boundary.

Concept of Stability

The state vector v of the system, Fig. 1, contains discrete and elastic variables. The system is described by the set of differential equations

$$\dot{v} = F\left(v, \frac{\partial v}{\partial y}, \ldots \frac{\partial v^4}{\partial y^4}\right) \qquad (1)$$

where F is a vector function. Among the many theorems on stability of the class developed by Liapunov and Krasovskii, two have direct relevance to the development here. Let $\bar{U}$ be a scalar functional depending on v and on spatial derivatives of v then the following theorems are used:

Theorem 1: If there exists for system (1) a positive (negative) definite scalar functional $\bar{U}$ whose total time derivative $d\bar{U}/dt$ is negative (positive) semidefinite along every trajectory of (1), then the trivial solution $v \equiv 0$ is stable (Liapunov).

Theorem 2: If the conditions of Theorem 1 are satisfied and if additionally the set of points at which $d\bar{U}/dt$ is zero, does not contain a nontrivial positive half-trajectory, then the trivial solution $v \equiv 0$ is asymptotically stable (Krasovskii).

In the following it will be shown that under certain circumstances the Hamiltonian of the given system constitutes an appropriate Liapunov functional. The requirement of Theorem 2, that energy dissipates for any possible motion other than the nominal motion in the neighbourhood of the nominal motion, is identical with the requirement of "complete damping". However, the damping of a freely spinning body with internal energy dissipation is not complete in terms of inertial attitude angles which are zero prior to perturbation, since after perturbation the vehicle cannot return to its original state. Thus, the asymptotic stability cannot be proclaimed as a consequence of the positive definiteness of the Hamiltonian. The method to circumvent this problem is to constrain the attitude angles through the angular momentum integral such that they represent deviations from an inertial direction Z coinciding with the instantaneous angular momentum vector $\underline{L}_G$ after perturbation from its nominal inertial orientation. The constrained attitude angles will assure complete damping and asymptotic stability.

The Hamiltonian

The Hamiltonian H can be expressed as the sum of kinetic and potential energy

$$H = T + V \quad . \tag{2}$$

The kinetic energy T of the system can be derived from the general expression

$$T = \frac{1}{2} \int\limits_{K_0, K_A} \overset{*}{\underline{R}}_s \overset{*}{\underline{R}}_s dm \tag{3}$$

where $\underline{R}_s$ is an inertial position vector of a mass element dm and K_0 and K_A denote the integration over the flexible beams K_I, K_{II} with the tip bodies K_1, K_2 and the rigid central body K_0. The asterisk above a vector denotes time differentiation with respect to an inertial frame. Since the system is assumed force-free, the mass center of the deformed body is inertially fixed. After some algebraic transformations the kinetic energy can be written conveniently in terms of matrix notation

$$T = \frac{1}{2}\,\underline{\omega}^T\,{}_G\underline{\underline{J}}\,\underline{\omega} + \underline{s}^T\underline{\omega} - \frac{1}{2}\,m\,\dot{\underline{r}}_G^T\dot{\underline{r}}_G$$

$$+ \frac{1}{2}\,\rho \int_{K_I,K_{II}} \dot{\underline{u}}^T\dot{\underline{u}}\,dy + \frac{1}{2}\,m_E\,(\dot{\underline{u}}^T\dot{\underline{u}})\Big|_{K_1,K_2} \qquad (4)$$

where the dot denotes time differentiation of that vector with respect to the body fixed frame P. The frame P coincides with the principal axes xyz of the body in the undeformed state. Further, $\underline{\omega}$ is the vector of the inertial angular velocity of the central body K_0, ${}_G\underline{\underline{J}}$ is the inertia dyadic of the deformed body referred to the frame G which is parallel to P but has the origin at the center of mass. The elements of the vector $\underline{s}$ depend on the elastic deformations and their time derivatives with respect to the body fixed system P, $\underline{r}_G$ is the vector of the position of the center of mass of the undeformed body with respect to the center of mass of the deformed body. The vector $\underline{u}$ contains the elastic displacements u_1 and u_3 of a mass element dm in the deformed state. The mass of the entire body K is m. The mass of the tip bodies K_1 resp. K_2 is m_E and ρ is the mass per unit length of the beam. The symbol $(\ )|_{K_1,K_2}$ denotes the boundary value at the location of the tip bodies K_1 and K_2.

For a force-free system the potential energy V arises primarily from the body elasticity. Special consideration has to be given in this case to the centrifugal forces. In Ref. (9) a detailed derivation of the potential energy is demonstrated. Thus, the potential energy is obtained in the form

$$V = \frac{1}{2} \int\limits_{K_I, K_{II}} (EJ_3 u''^2_1 + EJ_1 u''^2_3) dy$$

$$+ \frac{1}{2} \int\limits_{K_I, K_{II}} P(y)(u'^2_1 + u'^2_3) dy \tag{5}$$

where P(y) is the centrifugal force acting at the location y and EJ_1 resp. EJ_3 is the bending stiffness of the elastic beam. The prime over u_1 resp. u_3 denotes the spatial derivative.

The Force-Free System

For the force-free system the angular momentum vector $\underline{L}_G$ about the center of mass is defined as follows

$$\underline{L}_G = \int\limits_{K_0, K_A} \underline{r}_G \times \overset{x}{\underline{R}}_s dm = \text{const.} \tag{6}$$

where $\underline{r}_G$ is the position vector from the center of mass to the mass element dm. In matrix notation the angular momentum vector can be expressed as

$$\underline{L}_G = \underline{\underline{J}}_G \underline{\omega} + \underline{s} = C\Omega\underline{L} \tag{7}$$

where C is the principal moment of inertia about the z-axis in undeformed state, Ω is the spin angular velocity of the nominal motion and $\underline{L}$ is the vector of the direction cosines between the spin axis Z and the axes of the body fixed frame P. The vector $\underline{L}$ does not contain the generalized coordinate Θ_3 corresponding to the rotation about the z-axis. However, $\underline{L}$ can be expressed by the generalized coordinates Θ_1 and Θ_2, corresponding to the rotation about the axes x and y of the frame P relative to an inertial frame.

Equation (7) can be used to eliminate the angular velocity $\underline{\omega}$ in the kinetic energy (4). With equations (4), (5) and (7) and a series expansion neglecting the third order and higher terms in $\Theta_1, \Theta_2, u_1, u_3$ as well as constant terms and terms which are always positive by definition, after some partial integration, the following functional is obtained

$$W = \Omega^2 \{ \frac{C}{B} (C-B)\Theta_1^2 + \frac{C}{A} (C-A)\Theta_2^2$$

$$+ 2 \frac{C}{B} \Theta_1 \left[\rho \int_{K_I, K_{II}} u_3 y dy + m_E (u_3 y) \Big|_{K_1, K_2} \right]$$

$$- \rho \int_{K_I, K_{II}} u_1^2 dy - m_E u_1^2 \Big|_{K_1, K_2}$$

$$+ \frac{1}{B} \left[\rho \int_{K_I, K_{II}} u_3 y dy + m_E (u_3 y) \Big|_{K_1, K_2} \right]^2 \} + W_V \quad (8a)$$

where W_V is defined by

$$W_V = \int_{K_I, K_{II}} \{ \frac{1}{2} \left[EJ_3 u_1 u_1'''' + EJ_1 u_3 u_3'''' \right]$$

$$- \left[u_1 \frac{\partial}{\partial y} (P(y) u_1') + u_3 \frac{\partial}{\partial y} (P(y) u_3') \right] \} dy$$

$$- \left[EJ_3 u_1 u_1''' + EJ_1 u_3 u_3''' \right.$$

$$\left. - P(y)(u_1 u_1' + u_3 u_3') \right] \Big|_{K_1, K_2} \quad (8b)$$

and A, B and C are the principal moments of inertia of the undeformed body.

The problem of investigating the stability reduces to that of testing the expression (8) for definiteness. Circumventing this problem by modal analysis will now be considered.

Model Analysis of a Rotating Elastic Beam with Tip Body

To perform a stability analysis by means of the normal modes procedure, it can be assumed that the elastic displacements can be represented by infinite series of suitable space-dependent eigenfunctions $U_{(k)i}(y)$ multiplying corresponding time-dependent generalized coordinates $u_{ki}(t)$ in the following form

$$u_k = U_{(k)i}(y) \cdot u_{ki}(t) \qquad i = 1,2,3\ldots\infty, \qquad k = 1,3 \tag{9}$$

The eigenmotion in the meridian plane will be considered only. Similar results are obtained for the motion in the equator plane. The eigenfunction U satisfies the eigenvalue problem

$$L(U) = \Lambda^2 M(U) \tag{10}$$

and the boundary conditions

$$y = R \quad : \quad B_1(U) = 0 \quad , \quad B_2(U) = 0 \tag{11a}$$

$$y = R{+}L : \quad B_3(U) = 0 \quad , \quad B_4(U) = \Lambda^2 C_4(U) \tag{11b}$$

with the differential operators

$$L = EJ_1\frac{\partial^4}{\partial y^4} - \frac{\partial}{\partial y}\left(P(y)\frac{\partial}{\partial y}\right) \quad , \quad M = \rho \quad , \quad B_1 = 1 \quad ,$$

$$B_2 = \frac{\partial}{\partial y} \quad , \quad B_3 = \frac{\partial^2}{\partial y^2} \quad ,$$

$$B_4 = -EJ_1\frac{\partial^3}{\partial y^3} + \Omega^2 m_E(R+L)\frac{\partial}{\partial y} \quad , \quad C_4 = m_E \tag{12}$$

where Λ is the eigenfrequency corresponding to the eigenfunction U.

In Ref. (9) it is shown that in this case the eigenvalue problem is self-adjoint. For this purpose an extended definition of self-adjointment with respect to the eigenfrequency dependent dynamic boundary condition (B_4, C_4) has been introduced, by means of which the generalized orthonormality relation

$$\rho \int_R^{R+L} U_i U_j dy + m_E (U_i U_j)\Big|_{R+L} = \delta_{ij} \qquad i,j = 1,2,3,\ldots\infty \tag{13}$$

can be determined where δ_{ij} is the Kronecker symbol.

The numerical determination of the eigenfunctions is also complicated due to the fact that the dynamic boundary conditions depend on the eigenfrequencies. Thus, the usual definition of the Rayleigh quotient for continuous systems has to be extended to eigenfrequency dependent boundary conditions (see (9)).

Stability Criteria

Setting the symmetric and asymmetric mode

$$u_k = U_{(k)(s)i}(y) \cdot u_{ksi}(t) + U_{(k)(a)i}(y) u_{kai}(t) \qquad i = 1,2,\ldots\infty \quad k = 1,3 \tag{14}$$

with the relation

$$U_{(k)(s)i}(y) = U_{(k)(a)i}(y) = U_{(k)(s)i}(-y) = -U_{(k)(a)i}(-y) = U_{(k)i} \tag{15}$$

and the orthonormality properties (13) for the functional W the following function is obtained

$$W = W_s + W_a \tag{16a}$$

with

$$W_s = \Lambda^2_{(1)i} u_{1si} + \Lambda^2_{(3)} u_{esi} \tag{16b}$$

$$W_a = \frac{1}{2}\,\Omega^2 \left[\frac{C}{B}\,(C-B)\,\Theta_1^2 + \frac{C}{A}\,(C-A)\,\Theta_2^2\right] + 2\,\frac{C}{B}\,\Theta_1 \Omega^2 A_i u_{3ai} + 2\,\frac{1}{B}\,\Omega^2 A_i A_j u_{3ai} u_{3aj} + \Lambda^2_{(3)i} u^2_{3ai} + \Lambda^2_{(1)i} u^2_{1ai} \qquad i,j = 1,2,\ldots\infty \tag{16c}$$

where A_i is defined by

$$A_i = \rho \int_R^{R+L} U_{(3)i}\, y\,dy + m_E (R+L)\, U_{(3)i}\Big|_{R+L} \; . \tag{17}$$

For asymptotic stability, W has to be positive definite. W_s depends on the symmetric mode only and is always positive definite due to the positive definite eigenvalue problem. Therefore, W is surely positive definite if W_a is positive definite. After a congruence transformation (see Ref. 9) the positive definiteness can be obtained by means of the Sylvester-criterion as follows

$$k_1 < 0 \tag{18}$$

$$k_2 > 0 \tag{19}$$

$$k_1 < -2\,\frac{A_i^2}{A}\left(\frac{\Lambda_{(3)i}}{\Omega}\right)^{-2} \qquad i = 1,2,3,\ldots\infty \tag{20}$$

where $k_{1/2}$ are the inertia parameters defined by

$$k_1 = \frac{B-C}{A} \quad , \quad k_2 = \frac{C-A}{B} \; . \tag{21}$$

Due to the stability theorems 1 - 2 the stability criteria obtained are sufficient conditions for asymptotic stability of the nutation Θ_1, Θ_2 of the central body and the beam deformations u_1 and u_3. These stability regions are shown in the Magnus shape triangle (Fig. 2).

The inequalities (18) and (19) represent the well known rigid body criteria $k_1 k_2 < 0$ (vertically shaded region in the shape triangle) and the well known "energy sink" criteria $k_1 < 0$ and $k_2 > 0$ (horizontally shaded region in the shape triangle). Sufficient conditions are obtained by adding condition (20) (diagonally shaded region in the shape triangle). According to (20) the magnitude of this stability region is essentially depending on the "elastic" system properties.

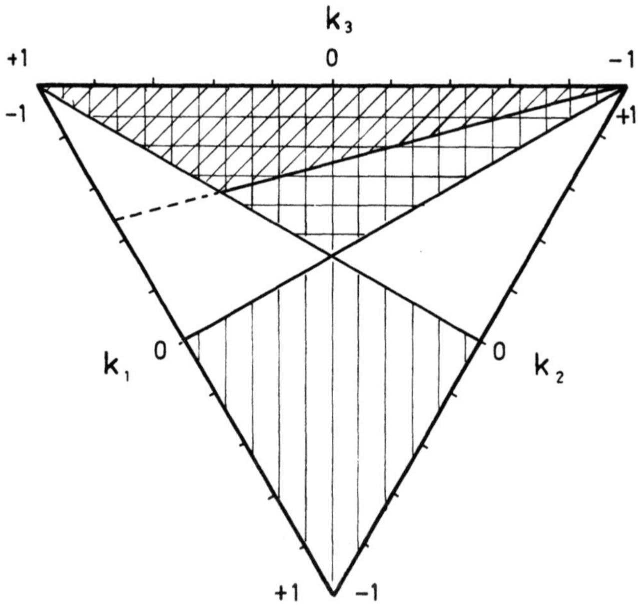

Figure 2. The Magnus shape triangle.

Numerical Results

For the numerical investigation the following dimensionless parameters are defined

$$e = \frac{EJ_1}{\rho L^4 \Omega^2} \quad , \quad a = \frac{m_E}{\rho L}$$

$$r = \frac{R}{L} \quad , \quad k_a = \frac{A_A}{A} \qquad (22)$$

where A_A is the moment of inertia of the body k_A in undeformed state about the axis x.

Figure 3 demonstrates the dependence of the stability boundary k_1 of the number N of the considered eigenfunctions. It will be realized that the stability boundaries converge with $N \to \infty$. Moreover, it is shown that the stability conditions for $N<\infty$ are not sufficient any longer. The error due to truncation depends

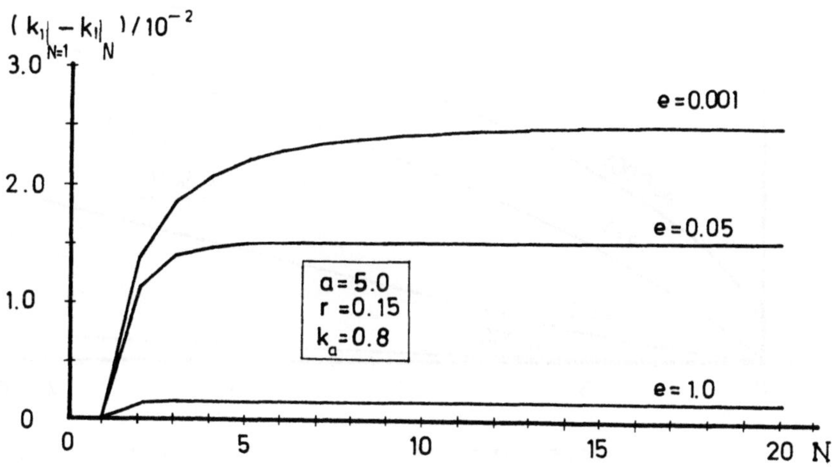

Figure 3. The dependence of the stability boundary of N.

on the difference between the estimated eigenfunctions and the true eigenfunctions as well as of the number of the estimated eigenfunctions which will be considered in stability criteria.

Figure 4 shows the dependence of the stability boundary k_1 (N=7) on the parameter a (mass of the tip body to the mass of the beam) for various values e. For a = 0 the stability boundaries without tip body are obtained. The boundary curve for e ≡ 0 is identical with the stability boundary of a string with tip body, the boundary for e → ∞ is identical with the stability boundary of a rigid body. Thus, the influence of the elasticity of the beam and the influence of the tip body on the stability is demonstrated. The stability boundary independence of the remaining system parameters is shown in Ref. (9).

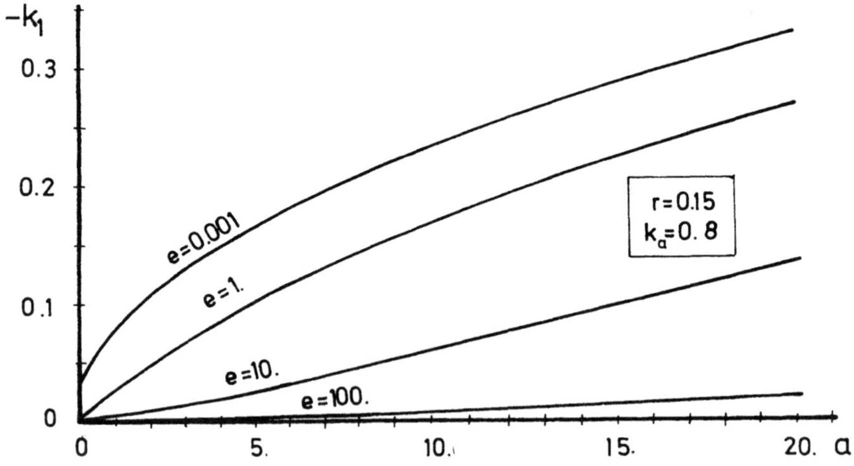

Figure 4. The dependence of the stability boundary on the system parameter.

References

1 Thomson, W.T.; Reiter, G.S.: Attitude Drift of Space Vehicles. J. Astron. Sci. n. 2, 7(1960) 29-34.

2 Meirovitch, L.: Attitude Stability of an Elastic Body of Revolution in Space. J. Astron. Sci. n. 4, 8 (1961) 110-113.

3 Pringle, R., Jr.: Bounds on the Vibration of a Symmetrical Satellite. AIAA Journal n. 5, 2 (1964) 908-912.

4 Likins, P.W.: Stability of a Symmetrical Satellite in Attitudes Fixed in an Orbiting Reference Frame. J. Astron. Sci., n. 1, 12 (1965) 18-24.

5 Wang, P.K.C.: Stability Analysis of a Simplified Flexible Vehicle via Liapunov's Direct Method. AIAA Journal, n. 9, 3 (1965) 1764-1766.

6 Parks, P.C.: A Stability Criterion for Panel Flutter via the Second Method of Liapunov. AIAA Journal, n. 9, 4 (1966) 175-177.

7 Meirovitch, L.; Calico, R.A.: The Stability of Motion of Satellites with Flexible Appendages. NASA CR-1978 (1972).

8 Likins, P.W.; Barbera, F.J.: Attitude Stability of Spinning Flexible Spacecraft. NASA Cr NAS8-26214 (1971).

9 Lohmeier, P.: Strukturanalyse and Stabilität Rotierender Elastischer Raumstationen. Diss. Institut B für Mechanik, Tech. Universität München (appears soon).

Non-Gravitational Forces in Satellite Dynamics

L. SEHNAL
Astronomical Institute of the
Czechoslovak Academy of Sciences.
Ondrejov, Czechoslovakia

Abstract

The effects of different non-gravitational forces on the motion of artificial satellites are described. Drag, direct solar radiation pressure and Earth albedo radiation pressure are discussed in detail. The effect of the Earth's shadow may be described analytically by different types of "shadow function". However, this approach can cause errors in practical computation of perturbations. The rotation of the body must be taken into account in some cases. The albedo radiation pressure can be described analytically provided the constant albedo model is used, but the practical computation requires numerical methods, owing to the character of the albedo distribution over the Earth surface.

Introduction

The non-gravitational forces, causing perturbations in the motion of artificial satellites, may be divided in principle into two main effects, atmospheric drag and radiation pressure, which may again be divided into the direct solar radiation pressure and the reradiation of the Earth, of which the albedo component is the most important part. The respective disturbing forces have as a common feature dependence on the size and physical properties of the satellite's surface. There exist certainly some other disturbing effects of non-gravitational origin, such as the electromagnetic forces, but they are much smaller and up till now their influence on the observed behavior of satellite motion has not been reliably proved.

Magnitudes and Limits of Disturbing Forces

The basic expressions describing the disturbing accelerations caused by the effects mentioned above are: (i) atmospheric drag:

$$F_D = - \frac{1}{2} C_D \frac{A}{m} \rho V^2 \quad , \tag{1}$$

where C_D is the drag coefficient, A/m is the satellite area-to-mass ratio, ρ is the density of the atmosphere and V is the satellite velocity with respect to the atmosphere; (ii) direct solar radiation pressure:

$$F_S = K \frac{A}{m} Q \quad , \tag{2}$$

where K is the reflectivity coefficient and Q is the radiation pressure exerted on a unit surface; (iii) Earth albedo radiation, which, under the most simple circumstances may be written as

$$F_A = \frac{1}{4} K \frac{A}{m} Q \alpha \left(\frac{R}{r}\right)^2 \quad , \tag{3}$$

where α is the albedo, R is the radius of the Earth and r is the radial distance of the satellite.

We note that the area-to-mass ratio is a common factor in these three equations and since it disappears when we compare the magnitudes of the disturbing accelerations, we can get the heights at which the radiation pressure effects exceed those of the drag. The limiting heights are given in case of direct solar radiation pressure by

$$\frac{\phi}{r} = \frac{2\ K\ Q}{C_D\ GM} \quad ,$$

where GM is the geocentric gravitational constant. In the case of albedo radiation pressure, the limits are given by

$$\rho r = \frac{K\ Q\ \alpha}{2\ C_D\ GM} \quad .$$

Using some atmospheric model, (e.g. Jacchia (1)), we may then find, for a given exospheric temperature (here 1400°K), that the effect of direct solar radiation pressure exceeds that of the drag at a height of 800 km and its magnitude on unit surface is of the order of 10^{-5} dyn/cm^2. The albedo radiation pressure becomes greater than the drag from heights of about 1200 km upward, being of the order of 10^{-6} dyn/cm^2. The height limits depend, of course, on the state of the atmosphere, determined by the solar activity.

The Atmospheric Drag

The mathematical description of the drag problem has already achieved high perfection and not much remains to be added. The most important factor in Eq. (1) is certainly the density of the atmosphere and since this changes according to the activity of the Sun sometimes very rapidly and unexpectedly, no precise computation can be made in an analytically continuous way.

For most of the computational work, the atmospheric density may be expressed as

$$\rho(z) = \rho_0(z)\ (1 + \beta \cos^n \frac{\psi}{2}) \quad ,$$

where z is the height above the Earth's surface, β and n are empirically determined coefficients and ψ is the angle between the radius-vector of the satellite and the axis of the atmospheric bulge (day-to-night variation). $\rho_0(z)$ means the atmospherical density, which changes with height according to

$$\log \rho_0(z) = a + b\,z + c\,\exp(dz) \quad , \qquad (5)$$

where a, b, c and d are again numerical coefficients, depending on the solar activity. The flattening of the atmosphere may be taken into account by considering the height of the satellite z to be given by

$$z = r - R\,(1 - f\,\sin^2\phi) \quad ,$$

where f is the Earth flattening and ϕ is the latitude. There may be some more elaborate descriptions of the atmospheric density; one of the last treatments by Fominov (2) uses a representation of the ellipticity of the atmosphere and its day-to-night variation by a series of Legendre polynomials. The results are, however, so complex that the author himself recommends the numerical solution of his equations.

Short-period Drag Perturbations

The coefficients in Eqs. (4) and (5) may be considered as constants over short intervals of time, certainly during one revolution of the satellite around the Earth, so that we can obtain the short-period perturbations of the orbital elements during one revolution (3,4). We see such perturbations in Fig. 1, where the effect of the atmospheric bulge (the day-to-night effect) is shown separately. The example chosen here (satellite 1958 γ) exhibits large secular changes after one revolution: 1.6 km in semi-major axis, since the perigee of the orbit lies very low, at 174 km.

We can see very clearly on the curves how the disturbing influence concentrates around the perigee, and this fact is being widely used in deriving the atmospheric density from the observed accelerations of satellites.

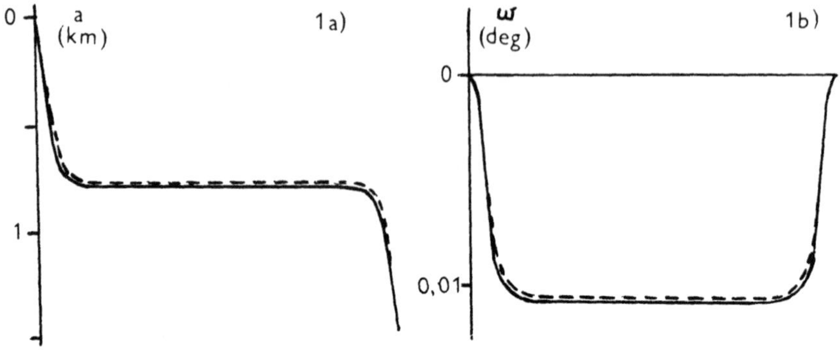

Figure 1. a) The short-period perturbations of the semi-major axis during one revolution. b) The short-period perturbations of the argument of perigee. The dashed lines denote the perturbations without considering the atmospheric bulge.

Long-term Drag Changes

After substituting the components of the disturbing acceleration derived from Eq. (1) into the Lagrangian equations for the changes of orbital elements, we see immediately the characteristic feature of the perturbations: since the disturbing force acts against the velocity vector, it changes mainly the shape and dimensions of the orbit, whereas the position of the orbit remains practically unchanged. The influence of drag on the semi-major axis and eccentricity causes the known contraction of the orbits, when the height of the apogee decreases much faster than that of perigee. In Fig. 2 we have the values of the perigee heights of satellite 1965 11 D, as given by King-Hele (5); these are the values corrected for the periodic effects of odd-zonal harmonics and of the luni-solar perturbations. The slow change during most of the satellite's life and the rapid decrease at the end is clearly seen.

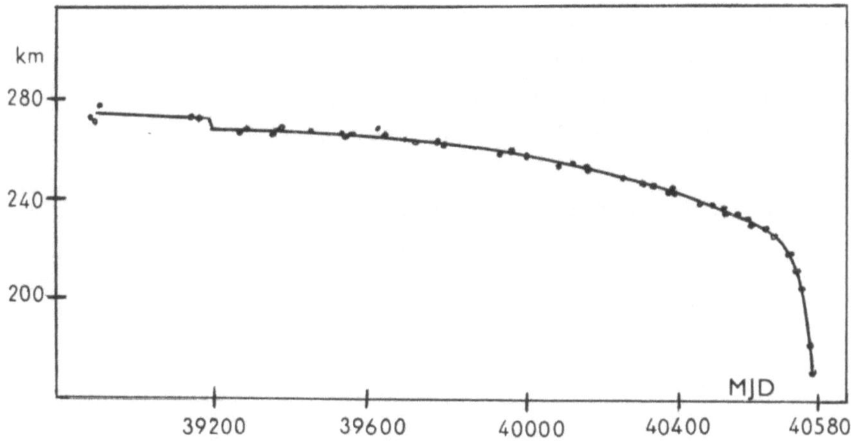

Figure 2. The perigee height of satellite 1965 11 D, corrected for odd-zonal harmonics and luni-solar perturbations (5).

The change of the semi-major axis causes changes of orbital period and since that quantity can be derived relatively precisely from the observations, it is the best known and most studied orbital characteristic connected with the drag effects.

The period of satellite revolution decreases secularly under the influence of the dissipative drag force, and, superimposed on the secular change, we can observe the periodic perturbations. They reflect the periodic changes of the density of the atmosphere, caused mainly by solar activity. Analyzing the mean motion of the satellites, we can detect the most important phenomena in the behavior of the atmosphere (6,7):

1. Variations caused by short-term changes in solar activity, as transferred to the Earth by streams of particles or by ultra-violet radiation. The changes are sometimes of random character but the 27-day period is usually clearly visible.

2. The semi-annual variation.
3. The day-to-night variation, which expresses itself in orbital perturbations of approximately the period with which the perigee revolves around the Earth.
4. The 11-year cycle variation.

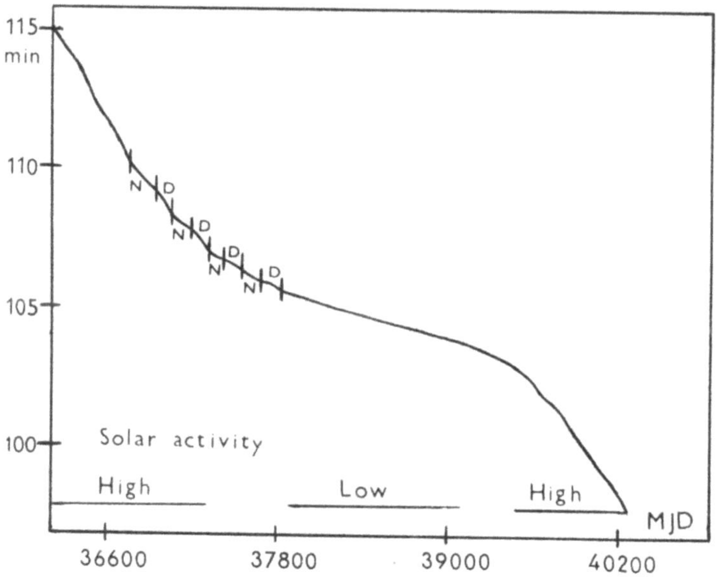

Figure 3. Orbital period of satellite 1958 α. The intervals with perigee in sunlight (D) and in darkness (N) are shown (6).

The orbital period of the satellite 1958 α in Fig. 3 shows some of those features very distinctly (6). The quantities characterizing the position of the orbit are affected by the atmospheric drag rather indirectly, as was pointed out by Jaeck and Barlier (8). Following their analysis, let us consider, for example, the equation for the change of Ω, caused by the asphericity of the Earth. This is

$$\frac{d\Omega}{dt} = \text{secular terms } (a,e,i) + \text{periodic terms}$$

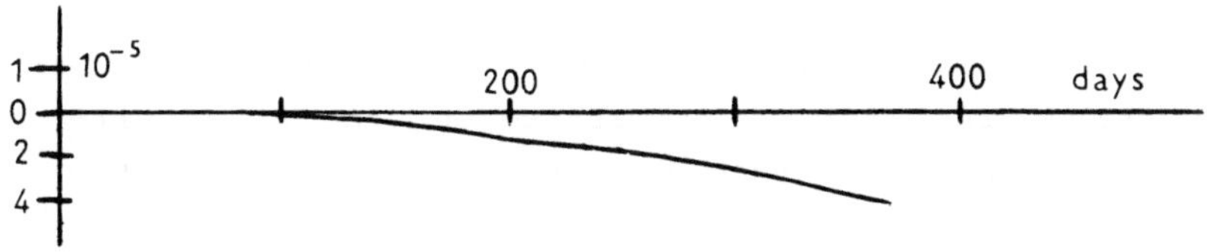

Figure 4. The mixed effects of drag and Earth's asphericity in Ω, as obtained by numerical integration (8).

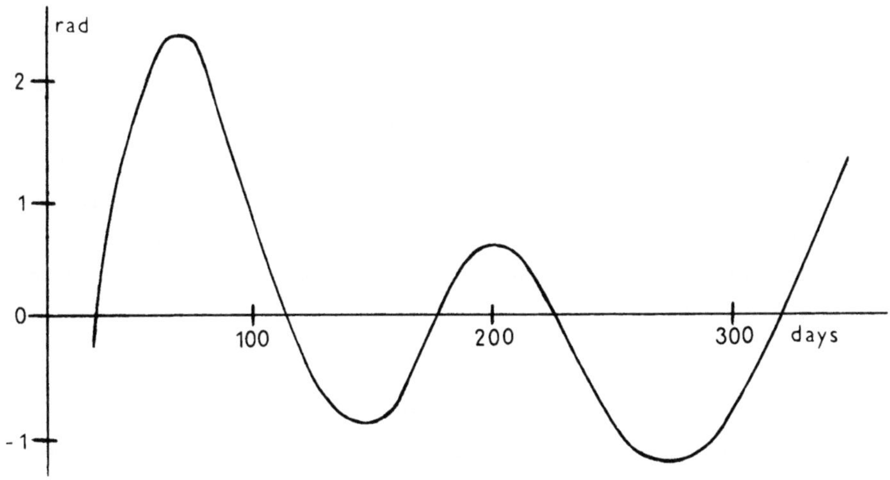

Figure 5. The periodic part of the perturbation of Ω, from Figure 4.

Now if we consider the influence of drag, a and e are no longer constants; they contain again both secular and periodic terms. Thus, after integration, we get immediately two kinds of mixed perturbations: quadratic terms in t^2 and periodic terms. The periodic terms have then amplitudes depending on the changes of semi-major axis, caused by the drag. The numerical integration, made by Jaeck and Barlier, shows both terms quite clearly. In Fig. 4 we have the change in Ω for the satellite Geos-A; after fitting a parabola to the curve, the residuals show the periodic part (Fig. 5).

The A/m and C_D Coefficients

The quantitative determination of the magnitudes of the drag effects requires also a knowledge of the factors A/m and C_D in Eq. (1).

The determination of the A/m ratio may cause difficulties in the case of uncontrolled tumbling of the satellite, and especially a body of spheroidal shape may produce rotation with precessing motion of the axis of rotation. Nevertheless, the A/m ratio value is usually the best known quantity of Eq. (1) and once determined, it may be considered as constant for drag studies (8,9).

The drag coefficient C_D is considered usually to be constant and equal to 2.2. But the detailed analyses (10,11) show that it changes quite distinctly with height, since it depends on the properties of the satellite's surface as this interacts with the surrounding medium. The coefficient C_D is usually determined theoretically, since laboratory measurements encounter great difficulties when imitating the conditions of molecular motions at great heights. The widely used value 2.2 seems now to be rather low even in the lower regions of the atmosphere (7). From the analyses of Cook (11) it follows that up to heights of 400 km this coefficient may be taken as constant, but that it then increases rapidly with increasing height. Kohler in (8) computed the value of C_D for the proposed satellite D 5 B and found a value of 2.8 at the height of 1400 km. In one of his papers, Slowey takes for C_D the value 3.6 at the height of 3300 km. Thus, neglecting the variation of C_D with height might lead to substantial errors, almost of a factor of 2.

The Rotation of the Atmosphere

The rotation of the atmosphere gives rise to a force perpendicular to the orbital plane and this force causes changes in inclination and in ascending node. These changes are very small and are perceptible only during longer time intervals, especially during the whole lifetime of the satellite, since at the end of the satellite's existence the changes are more pronounced. Since the position of the ascending node suffers great perturbations from other disturbing forces, the effect of the atmospheric rotation is observed mainly in changes in the inclination. From the analyses of the inclinations of many satellites, mostly by King-Hele (12), it has been found that the rotation speed of the atmosphere increases up to a height of approximately 400 km, where it is about a factor 1.4 greater than the rotation speed of the Earth, and then decreases again.

In Fig. 6, we see the inclination of the satellite 1971 104 1, the full line denoting the theoretical value of the atmospheric rotation effect (13). The initial value of the inclination and that of the ratio of the speed of atmospheric rotation to that of the Earth are chosen as initial constants to make the

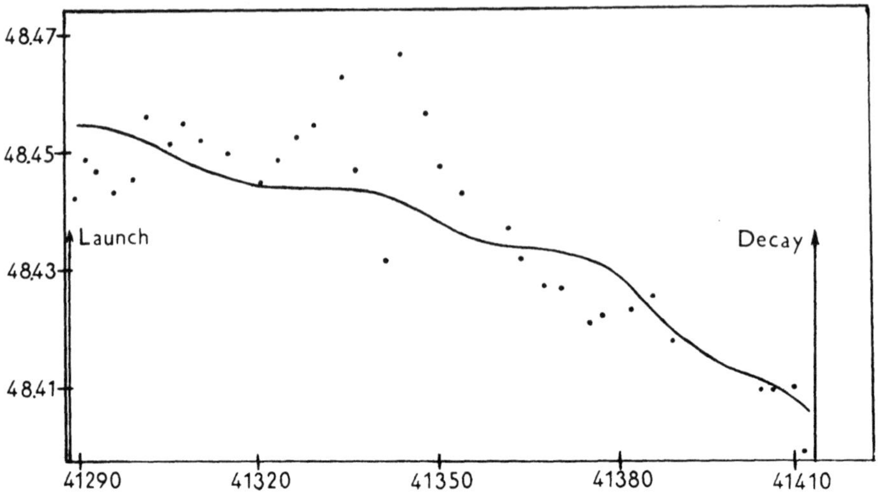

Figure 6. The inclination of satellite 1971 104 1 for a ratio of the speed of atmospheric rotation to that of the Earth of 1.3: full line, theoretical value; dots, observed values; vertical numbers, degrees; horizontal numbers, days.

best fit of the theoretical curve to the observed values. The model of the atmosphere includes only the flattening; this is certainly sufficient from the point of view of the precision of the orbital elements. The atmospheric model might include also the atmospheric bulge (day-to-night variation) (14). In that case we might have sometimes a 25% increase of the effect, but since this would be for a short interval of time only, it would be very difficult to observe. As pointed out by King-Hele (14), there might appear also the effect of resonance under some favorable conditions. But again this effect is more of theoretical interest than a really observable value.

Electrically Charged Satellite

The artificial satellites move in a region which one may consider as an ionized plasma, so that by interaction with the free electrons and ions a satellite acquires an electrical charge. Then, in essence, two effects take place: first, disturbing forces arise of a character similar to that of atmospheric drag, i.e. they act against the motion of the satellite, and secondly, a Lorentz force arises, due to the motion of an electrically charged body in the magnetic field of the Earth. The drag charge may arise for various reasons: a Coulomb drag which causes an increase in satellite cross-section area (15), an induction drag, which arises from the interaction of electrical currents in the satellite with the Earth's magnetic field (16), and a wave drag, caused by the plasma charge separation by a moving satellite (17,18).

The effects generally increase with increasing height and depend on the dimensions of the satellite. The first estimates of their magnitudes were somewhat high; so far there has been no evidence of any charged drag up to heights of about 700 km. In 1969, Fea (19) suggested the presence of induction drag at heights of about 2400 km, in an attempt to explain the changes in the orbital accelerations of satellite 1963 30 D, but this was later completely explained by Slowey (20), by the variable albedo effect. Thus, even when analyzing the motion of large satellites at great heights, no evidence was found of the charged drag effect.

The same may be said about the Lorentz force. A study (21) which looked for the explanation of the rapid decrease of the satellite orbital inclination led to the conclusion that a satellite must have an electrical potential of at least 1000 volts in order to show any observable effect. Such a high value was, of course, never observed, nor is it predictable theoretically; the measured and computed values agree on a potential of about 3 orders lower.

Solar Radiation Pressure

The effects of solar radiation pressure are already very well known; their more detailed study began with the launching of the Echo balloon satellites. However, the investigations of this problem are in most cases made from a rather passive point of view, where it is not our aim to improve the knowledge of the basic physical phenomenon causing the effect, as it is in the case of the Earth's gravitational field or the atmosphere, but rather to analyze the observed motion of the satellite: we want to exclude these effects as an unnecessary burden to other investigations. Nevertheless, the knowledge of the phenomenon gives us an opportunity to take advantage of its consequences in order to choose some special orbits, so that we may make proper use of the resonance effect. The project West Ford is probably the best known purposeful usage of solar radiation pressure in satellite dynamics.

The theory of the problem was treated from very many angles (22) and the resonance effects gave a good field for detailed theoretical studies (23-25) and also for practical applications (26). In one of the most recent papers, Polyachova (27) considers, for example, the secular effects over long time intervals and represents the solar radiation pressure by the potential of a ring of negative mass around the Earth. The method uses a well known Gaussian (28) analysis of the planetary secular perturbations and permits the treatment of both radiation and gravitational effects together.

In the following, we shall discuss in more detail several special problems of the radiation pressure effects which have been recently studied.

The Shadowing Effect

When treated mathematically the shadowing effect causes some obstacles to analytical theories. During one revolution of the satellite, it is possible to derive the perturbations by computing precisely the moments at which the satellite leaves and enters the shadow and then integrate between them. This is a widely used procedure, described already, for example, by Kozai (29). To find the shadow limits in such a way requires, however, the solution of a quartic which cannot be simply expressed in terms of orbital elements, so that an analytical solution of the problem is not possible in that way.

Since the first treatment of the effects of the shadow, several attempts have been made to establish a so-called "shadow function", an analytical expression which will be equal to unity when the satellite is illuminated by the Sun and which vanishes when the satellite enters the shadow. Such a discrete function may be analytically represented by a Fourier series (30): however, in using this theory in practical computation one must take a large number of terms in order to achieve sufficient accuracy.

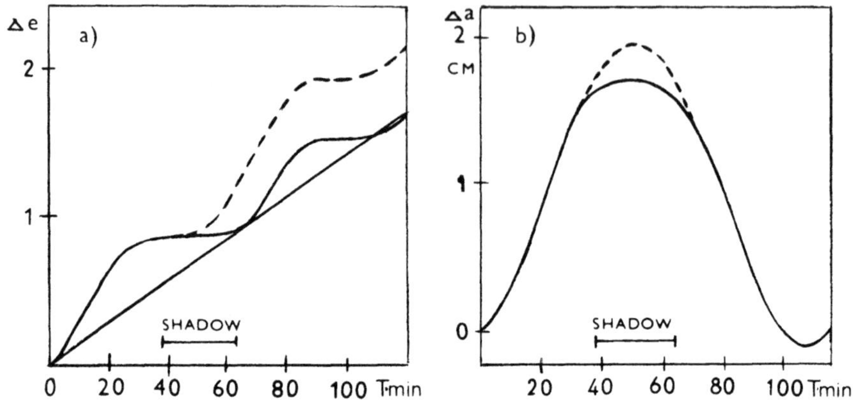

Figure 7. a) The perturbations of the eccentricity of the satellite Echo 1. b) The perturbations of the semi-major axis of the satellite Echo 1. The dashed lines show the results when the shadowing effect is not taken into consideration.

A practically usable shadow function was found by Lāla and Sehnal (31); it is a power series development of the ratio of a natural and absolute value of a quantity, which is positive when the satellite is in sunlight and negative when the satellite is in shadow. As an example of how it works, we have in Fig. 7 the perturbations of some elements of the orbit of the satellite Echo 1.

Another representation of the shadow function appeared lately in the work of Vashkoviak (32), who develops the function in a series of Legendre polynomials.

The practical use of a shadow function is, of course, limited by the number of terms we need to take into account in the development. Quite recently, Gaposhkin (33) tested the long-term prediction of the radiation pressure effect on an orbit of a hypothetical satellite, using first the numerical integration procedure and second the analytical theory with the shadow function. In the numerical integration the limits were computed from a quartic, whereas the shadow function was represented by a development in a power series (31). The results are compared in Fig. 8, for the case of the inclination.

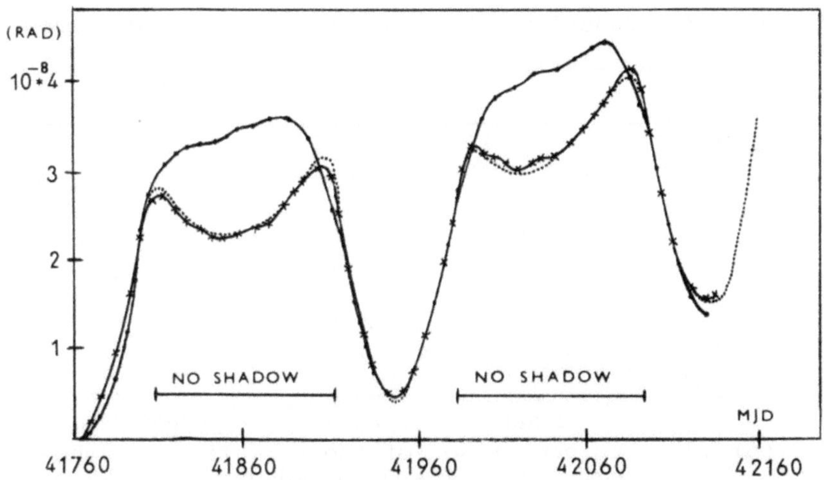

Figure 8. Comparison of the values of the inclination of a hypothetical orbit, computed by numerical integration (crosses), and analytically using either the shadow function (points) or computed points of intersection (dotted line).

We see that the two methods agree well when the satellite is entering the shadow, but differ when the satellite is in sunlight. The reason is that the shadow function does not give sufficiently good approximation to reality; especially when the satellite orbit approaches the shadow cylinder the shadow function is no longer equal to one and the effect is as though the satellite had already entered the shadow.

We see also how the curves agree with the results of an analytical theory, when proceeding step by step in computing the shadow limits precisely. In Fig. 9 we see another example of such a difference, when the points of intersection are computed or represented by the shadow function. The imperfection of the shadow function is clearly seen.

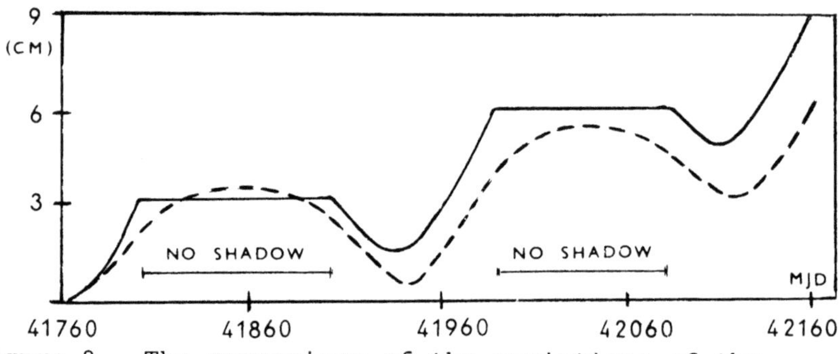

Figure 9. The comparison of the variations of the semi-major axis computed analytically by exact determination of the points of intersection (full line) and by using the shadow function (dashed line).

The comparison presented here was made only between two models of shadow limits, one given by the shadow function and the other given by exact determination of the points of intersection of the orbit with the shadow cylinder. However, neither of these models is precise; the penetration into the shadow should be considered not as a sharp limit: the spreading of the shadow boundary by the shadow function (Fig. 10) may be more similar to reality.

The shadow function considered here was only of one certain type, the power series development. But even if we use some other representation of the shadow function, I do not believe we should obtain any more satisfactory results.

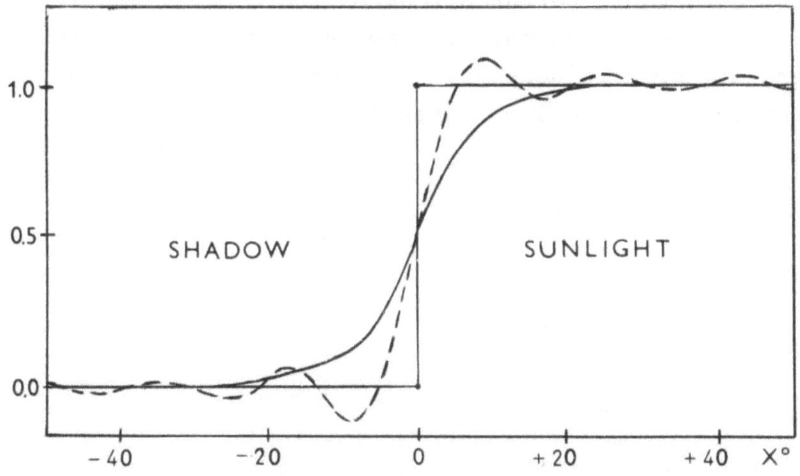

Figure 10. Power series (full line) and Fourier (dashed) line) development of the shadow function, as a function of geocentric angular distance from the point of intersection X♀.

Mixed Gravitational and Radiation Pressure Effects

The theory of radiation pressure effects shows clearly that there are no secular or long-period changes of semi-major axis and no second-order secular terms (34). However, these theories do not take into account the secular perturbations caused by the J_2 term in the Earth's gravitational potential. These perturbations are so great that they change the original character of the perturbations and, taken into account, they immediately give rise to the long-period changes of the semi-major axis, even if there is no shadowing effect. This was lately studied by Lāla (35) and from his investigations we may see in Figs. 11a and 11b these mixed perturbations transformed to the better observable perturbations of the orbital acceleration. The effect may be important when determining the atmospheric densities at greater heights from the motion of balloon satellites.

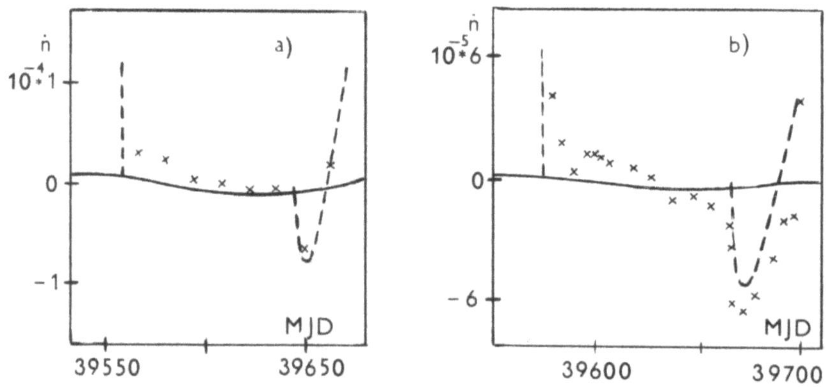

Figure 11. Comparison of computed and observed changes of the acceleration of satellite Pageos a) and Dash 2 b). The full line corresponds to the mixed gravitational and radiation pressure effects, the dashed line to the shadowing effect. Crosses are observed values. The Earth albedo radiation pressure was not taken into account.

Transversal Component of the Radiation Pressure

The observations of the changes in orbital periods of satellites 1966 56 A (Pageos) and 1963 30 D (Dash 2) showed that when the satellite penetrated the shadow periodic fluctuations arose, superposed on the main variation caused by the direct solar radiation pressure. The photometric observations of satellite Pageos by Vandenburgh and Kissel (36) showed that Pageos is not a sphere, but a prolate spheroid, the rotation axis of which precesses around the Earth-Sun line. This implied a presence of a transverse force, which might have great significance when acting in a discontinuous manner over the orbit, as shown by Smith and Kissel (37). The theory then showed an agreement to better than 10% with the observations. As may be seen in Fig. 12 (from Ref. 37), the amplitudes of the periodic changes reach almost the amplitude of the main perturbation itself.

Another recent study of this effect was made by Slowey (20), who investigated the orbital behavior of the satellite Dash 2, computing the changes of orbital elements by numerical integration and comparing the results with the observed values. He found excellent agreement of the theory with observations by introducing a transverse component of the disturbing force, without actually investigating the shape and position of the body of the satellite. In Fig. 13 we can see his comparison of theoretical and observed values of the acceleration of the satellite; the dashed curve shows the insufficiency of considering the main component of the disturbing force only.

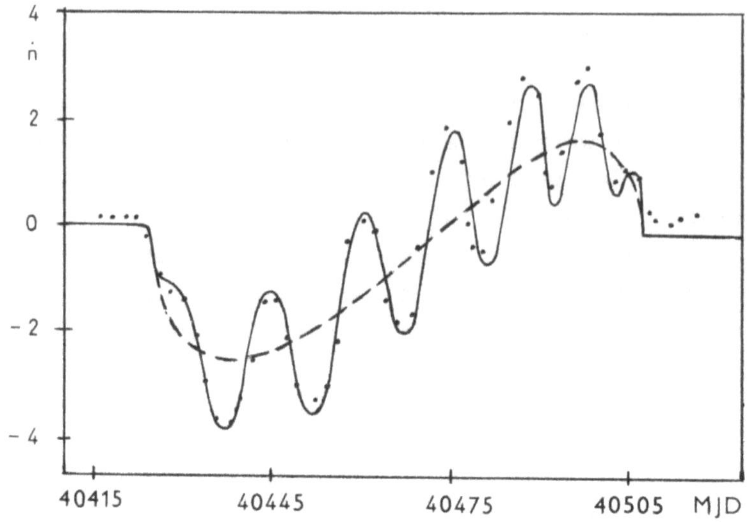

Figure 12. Observed and computed values of Pageos acceleration (Smith and Kissel (37)). The dashed curve corresponds to the main solar radiation pressure effect, dots indicate the observed values.

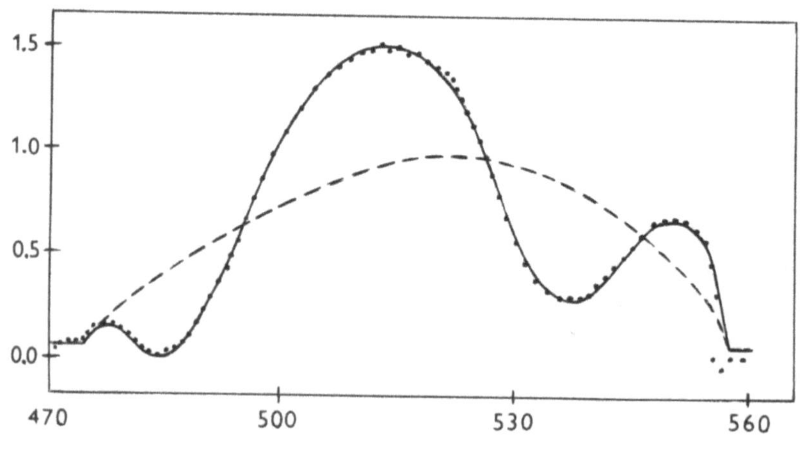

Figure 13. The orbital acceleration of satellite Dash 2 (dots are observed values). The full curve indicates the theoretical values and the dashed curve corresponds to the theoretical value without considering the transverse force. From Slowey (20). Vertical numbers, $-10^6 \dot{P}$, horizontal numbers, MJD.

As Slowey shows, the magnitude of the transverse force depends on the ratio of sphericity of the satellite body and on the angle of incidence of the radiation. When the angle is about 45º, the transverse component may reach 20% of the total force.

Earth Albedo Radiation Pressure

This effect was very much discussed years ago especially in connection with the Pageos satellite (19, 38). The analytical description of the effect is very difficult even when we take into account the simplest model of the Earth albedo - the constant value, homogeneously distributed over the Earth's surface (39,40). In this simple case it is possible to find an analytical expression which agrees with the numerical model in its limiting (maximum and minimum) values (vanishes when the satellite enters the Earth shadow). Such a function can be expressed, for example, as

$$F = F_C \cos^2\eta \, \frac{\pi/2}{\pi/2 + \cos^{-1} R/r}$$

where η is the angle between the satellite radius vector and the Earth-Sun line, R is the radius of the Earth. The quantity F_C is the maximum value of the albedo radiation pressure which can be computed see e.g. (41,42).

The albedo radiation pressure has its main component in the radial direction and may reach there a value of 1×10^{-5} dyn/cm^2 for a close satellite (500 km) and drop to 0.5×10^{-5} dyn/cm^2 for higher satellites (3000 km). The transverse component, if present, is at least one order lower.

At the last Symposium on the Dynamics of Satellites, in 1969, Prior (43) compared the theoretical and observed changes of eccentricity of Pageos satellite and obtained at least a qualitative agreement. Since it was then immediately found that the change of orbital period would be an excellent indicator of this effect, the accelerations of Pageos and Dash 2 satellites were used in the analysis (44).

The main difficulty which we encounter when computing the effect arises from the model of the albedo. Much has been said about the value of the average albedo; however, the Tiros 7 measurements showed its strong dependence on latitude and season. Then Slowey (44) computed the acceleration of the satellite Dash 2 using first the constant albedo model and then the Tiros albedo model. We may see from his Figs. 14a, 14b, that the differences caused by different models reach the order of the effect itself.

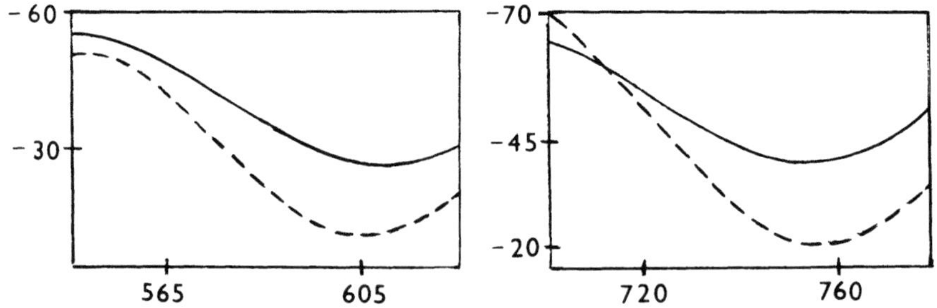

Figure 14. a), b) Orbital accelerations of Dash 2. The full curve corresponds to the Tiros model, the dashed curve to the constant albedo model.
a) Vertical numbers, $10^6\ \dot{n}$; horizontal, days.
b) Vertical numbers, rev/day^2; horizontal, days.

References

1 Jacchia, L.G.: Revised static models of the thermosphere and exosphere with empirical temperature profiles. SAO Spec. Rep. n. 332 (1971).

2 Fominov, A.M.: Secular and long periodic perturbations of the first order in the motion of artificial satellites due to the atmospheric drag. Bull. Inst. Theor. Astr. Leningrad, 13 (1973) 422-428.

3 Izaak, I.G.: Periodic drag perturbations of artificial satellites. Astr. Journal 65 (1960) 355-357.

4 Sehnal, L.; Mills, S.B.: The short-periodic drag perturbations of the orbits of artificial satellites. SAO Spec. Rep. n. 223 (1966).

5 King-Hele, D.G.; Walker, D.M.C.: Analysis of the orbit of 1965-11D (Cosmos 54 rocket). Planet. Space Sci. 21 (1973) 1081-1108.

6 King-Hele, D.G.: The upper atmosphere and its influence on satellite orbits. Dynamics of satellites, ed. by B. Morando, Springer 1970, 249-261.

7 Roemer, M.; Volland, H.: Die Hochatmosphäre im Flugbereich der Satelliten. Bundesministerium für wisschenschaftliche Forschung, Forschungsbericht W 68-01 (1968).

8 Jaeck-Berger, C.; Barlier, F.: Review of drag effects on satellite orbits for geodynamic studies. Proc. of First International Symposium on the Use of artificial satellites for geodesy and geodynamics, Athens 1973 (in press).

9 King-Hele, D.G.: Theory of satellite orbits in an atmosphere. London: Butterworths 1964.

10 Cook, G.E.: Satellite drag coefficients. Planet. Space Sci. 13 (1965) 929-946.

11 Cook, G.E.: Drag coefficients of spherical satellites. Ann. Géophys. 22 (1966) 53-64.

12 King-Hele, D.G.: The Rotational Speed of the upper atmosphere determined from changes in satellite orbits. Planet. Space Sci. 12 (1964) 835-853.

13. Sehnal, L.: Rotational speed of the atmosphere from observations of INTERKOSMOS satellites. Observ. of art. sat. of the Earth (in Russian), n. 12, Praha 1973, 257-262.

14 King-Hele, D.G.; Walker, D.M.C.: The change in satellite orbital inclination caused by the rotating atmosphere with day-to-night density variation. Celes. Mech. 5 (1972) 41-54.

15 Hohl, F.; Wood, G.P.: The electrostatic and electromagnetic drag forces on a spherical satellite in rarefied partially ionized atmosphere. Rarefied Gas Dynamics, ed. by J.A. Lautman (1963) 45-64.

16 Beard, D.B.; Johnson, F.S.: Charge and magnetic field interactions with satellites. J. Geoph. Res. 65 (1960) 1-7.

17 Chopra, K.P.: Interactions of rapidly moving bodies in terrestrial atmosphere. Rev. Mod. Physics 33 (1961) 153-189.

18 Drell, S.D.; Foley, H.M.; Ruderman, M.A.: Drag propulsion of large satellites in the ionosphere. J. Geoph. Res. 70 (1965) 3131-3145.

19 Fea, K.: The orbital acceleration of high balloon satellites. Dynamics of satellites, ed. by B. Morando, Springer 1970, 295-302.

20 Slowey, J.W.: Radiation pressure and air drag effects on the orbit of the balloon satellite 1963 30 D. SAO Spec. Rep. n. 356 (1974).

21 Sehnal, L.: The motion of a charged satellite in the Earth's magnetic field. SAO Spec. Rep. n. 271 (1969).

22 Sehnal, L.: Radiation pressure effects in the motion of artificial satellites. Dynamics of satellites, ed. by B. Morando. Springer 1970.

23 Hori, G.I.: The effect of radiation pressure on the motion of an artificial satellite. Space Math. Part. 3, Amer. Math. Sec. (1966) 167-178.

24 Polyachova, E.N.: Radiation pressure and the motion of Earth's satellites. Bull. Inst. Theor. Astr. Leningrad 9 (1963) 15-45.

25 Musen, P.: The influence of the solar radiation pressure on the motion of an artificial satellite. J. Geophys. Res. 65 (1960) 1391.

26 Shapiro, I.I.: Last of the West Ford dipoles. Science 154 (1966) 1445.

27 Polyachova, E.N.: An application of Gauss' method to the determination of secular radiational perturbations of Earth artificial satellites. Bull. Inst. Theor. Astr. Leningrad 13 (1972) 308-317.

28 Gauss, C.F.: Werke, Vol. 11 (1929).

29 Kozai, Y.: Effect of solar radiation pressure on the motion of an artificial satellite. SAO Spec. Rep. n. 56 (1961).

30 Ferraz-Mello, S.: Sur le problème de la pression de radiation dans la théorie des satellites artificiels. C.R. Acad. Sc. Paris 258 (1964) 463-466.

31 Lála, P.; Sehnal, L.: The Earth shadowing effects in the short-periodic perturbations of satellite orbits. Bull. of the Astr. Inst. Czechosl. 20 (1969) 327-330.

32 Vashkoviak, S.N.: The influence of radiation pressure on the motions of artificial satellites. Presented at the Symposium on motions of artificial satellites, Belogradčik, Bulgaria 1973.

33 Gaposhkin, E.M.: Private communication.

34 Ferraz-Mello, S.: Analytical study of the Earth's shadowing effects on satellite orbits. Celes. Mech. 5 (1972) 80-101.

35 Lála, P.: Combined gravitational and solar radiation pressure effects on the semi-major axis of the Earth satellite. Bull. Astr. Inst. Czechosl. 23 (1972) 342-345.

36 Vandenburgh, R.C.; Kissel, K.E.: Measurements of deformations and spin dynamics of the Pageos balloon satellite by photoelectrical photometry. Planet. Space Sci. 19 (1971) 223-231.

37 Smith, D.E.; Kissel, K.E.: Anomalous accelerations of the Pageos spacecraft. Space Research XII, Akademie Verlag, Berlin, GDR, 1972.

38 Smith, D.E.: Earth-reflected radiation pressure. Dynamics of Satellites, ed. by B. Morando, Springer 1970.

39 Sehnal, L.: The perturbations of the orbital elements caused by the pressure of the radiation reflected from the Earth. Trajectories of Artificial Celestial Bodies, ed. by J. Kovalevsky, Springer 1966.

40 Wyatt, S.P.: The effect of terrestrial radiation pressure on satellite orbits. Dynamics of satellites. ed. by M. Roy, Springer 1963.

41 Cunningham, F.C.: Earth reflected solar radiation input to spherical satellites. ARS Journal 32 (1962) 1033-1036.

42 Baker, R.M.L., Jr.: Radiation on a satellite in a presence of partly diffuse and partly specular reflecting body. Trajectories of artificial celestial bodies, ed. by J. Kovalevsky, Springer 1966.

43 Prior, E.J.: Observed effects of Earth reflected radiation and hydrogen drag on the orbital energies of balloon satellites. The use of artificial satellites for geodesy, ed. by S.W. Henriksen, Washington 1972.

44 Slowey, J.W.: Earth radiation pressure and the determination of density from atmospheric drag. Space Research XIV, 1974 (in press).

Solar Radiation Pressure and Balloon Type Artificial Satellites

R. VILHENA DE MORAES
Aeronautics Institute of Technology, Astronomical Observatory, São José dos Campos, S.P., Brazil

Abstract

The analytical model proposed by Ferraz Mello for the study of the effects of direct solar radiation pressure is used in order to compute the very long term perturbations for some balloon satellites (VANGUARD I, PAGEOS). In this model earth shadowing is taken into account. The results agree very well with those obtained from some direct numerical integrations. It is shown that this model can be suitably combined with the Brouwer-Hori model in order to study the long term perturbations on the trajectories of balloon satellites arising from the joint action of solar radiation pressure and atmospheric drag.

Introduction

The analytical study of the effects of the solar radiation pressure on artificial satellites, presents a difficulty due to the existence of a region where the disturbing effect ceases. This difficulty was overcome by Ferraz Mello (1) with the introduction of a "shadow function", suitably constructed, and that can be put into the form

$$\psi = 1 - \frac{\phi}{\pi} - \sum_{p \varepsilon N} \sin p\phi \cdot \cos p\lambda \qquad (1)$$

where

$$\phi = \arc\sin \left\{ \frac{R(1+e \cos v)}{a(1-e)^2} \right\} = \sum_{n=0} \phi_n \cos nv \qquad (2)$$

$$\cos \lambda = \sum_{i=1}^{6} A_i \cos(v+\lambda_i) \quad . \qquad (3)$$

Here ϕ_n depends only on a, e and on the Earth radius R, A_i (i = 1,2,...,6) is a function of the inclination of the orbit as well as of the obliquity of the ecliptic ε, and λ_i(i = 1,2,...6) depends on the Delaunay variables g, h and on the mean longitude of the Sun.

The integration of the equations of the motion with the shadow function included is extremely laborious. When one intends to use the computer in order to obtain explicitly the solution, it is necessary for the elaboration of a library of special subroutines that perform semi-analytically additions, products, integration and differentiations of Fourier series of several arguments as well as the introduction of the Cayley's Tables in a suitable step of the program. The number of terms generated by the considered series is very large. It is thus necessary to give a criterion for the elimination of those that do not give significant contributions to the final results.

The representation of the series as data for the machine and the position in the computer storage is fundamentally important to optimize the number of storage positions and the time of processing.

Following the ideas of Kovalevsky (2) those difficulties were overcome and the SRP program, for solving the equations proposed by Ferraz Mello, was elaborated in Fortran IV Language for IBM 1130 computer.

The Program

The SRP program is divided into 3 fundamental parts (3 connected programs):

SRP3A : Computation of the shadow function ψ.
SRP3B : Computation of the derivatives of the disturbing function with respect to the Delaunay elements.
SRP3C : Integration and output of the results.
Those programs are controlled by the SRP3X program in which the data input is presented.

The flow chart of the SRP3A program, when one limits N to NP is shown in Fig. 1.

Some Results

Initially the SRP program was applied to compute the long-period perturbations of the orbital elements of Vanguard II with NP = 6 for the period 36518-36638 (Modified Julian Days) and the results agree very well with those obtained by direct numerical integration, as may be seen in Fig. 2. The semi-major axis is expressed in earth equatorial radii. Figure 3 shows variation of the eccentricity and orbital inclination and the results are compared with those given by Kozai (3). The long-period variations of the semi-major axis of PAGEOS satellite were calculated also (NP=6), for comparison with the integration made by Lãla (4), Fig. 4. However, due to a bad choice of the initial orbital elements that introduced a near commensurability of the motion of the perigee and of the orbital plane with respect to the sun, this comparison is not valid for more than about 180 days.

The Effect of Drag

One of the effects of solar radiation pressure on balloon type satellites is a great variation in perigee's height which drives the satellite to lower atmospheric layers, where the effect of the air drag is stronger. A theory where these two phenomena (solar radiation pressure and air drag) are coupled is of a fundamental importance for the study of lifetimes of Earth satellites, and a combined analytical-numerical approach can be more advantageous than approximate analytical solutions, or numerical integrations.

If we couple Ferraz Mello's and Brouwer-Hori's theory (5) the equations of motion may be written as

$$\frac{dL_j}{dt} = \frac{\partial F}{\partial \rho_j} + P_j \quad ; \quad \frac{d\rho_j}{dt} = \frac{\partial F}{\partial L_j} - Q_j \quad ; \qquad (4)$$

$$F = \frac{\mu^2}{2L_1^2} - \frac{\mu}{r} + U$$

with

$$P_j = \sum_k \left(X_k + \psi \frac{\partial R}{\partial x_k} \right) \frac{\partial x_k}{\partial \rho_j} \quad ;$$

$$Q_j = \sum_k \left(X_k + \psi \frac{\partial R}{\partial x_k} \right) \frac{\partial x_k}{\partial L_j} \qquad (5).$$

where L_j, ρ_j ($j = 1,2,3$) are used for Delaunay variables. The force function U of the non-disturbed motion includes the main secular terms due the oblateness. X_k represents the drag acceleration, R is the force function of the solar pressure radiation and ψ is the shadow function.

In order to integrate the Eqs. (4) we must remember that we are really interested in a coupling effect. Thus, it is very important, that we preserve in the solution those terms Brouwer and Hori preserved in their work.

With this in mind, before the integration, (that can be made by the method of successive approximation since the right member of the equations will be suitably developed) we perform a transformation $(L_j,\rho_j) \rightarrow (L_j^*,\rho_j^*)$ where L_j,ρ_j are solutions of the problem without drag. Of course, if this transformation were a canonical transformation, the work done would be greatly reduced. But being less ambitious we can seek a transformation satisfying the following conditions:

a) $F(L_j,\rho_j) = F^*(L_j(L_j^*,\rho_j^*),\ \rho_j(L_j^*,\rho_j^*))$;

b) The difference among the old and the new variables must have the drag coefficient (at least, if we suppose that it is the smaller) as a factor.

Taking these hypotheses the second member, after the transformation, must be developed in terms of the eccentricity e with the use of the solution of the problem when we do not consider the drag effect (of course, if we want coupled terms, we need a 2nd order solution for that problem).

The developments are being done now with the aid of the SRP program subroutines.

Acknowledgments

I am very indebted to Dr. Ferraz Mello for his assistance in this work, and to my undergraduate student L. E. Penna Fernandes for his most valuable suggestions concerning my computer program. This research was partially supported by the National Council of Research under Contract No. 14211/71.

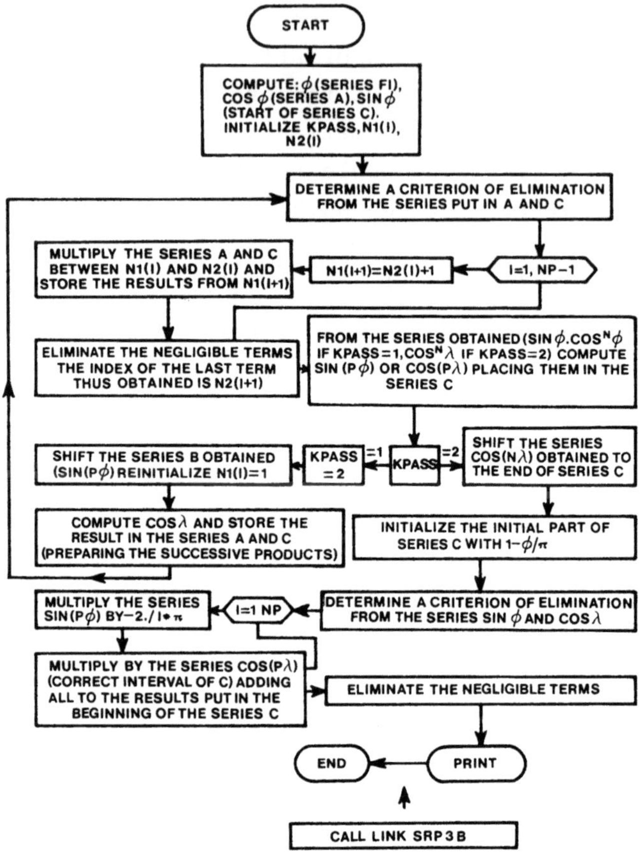

Figure 1. Flow chart of the SRP3A program.

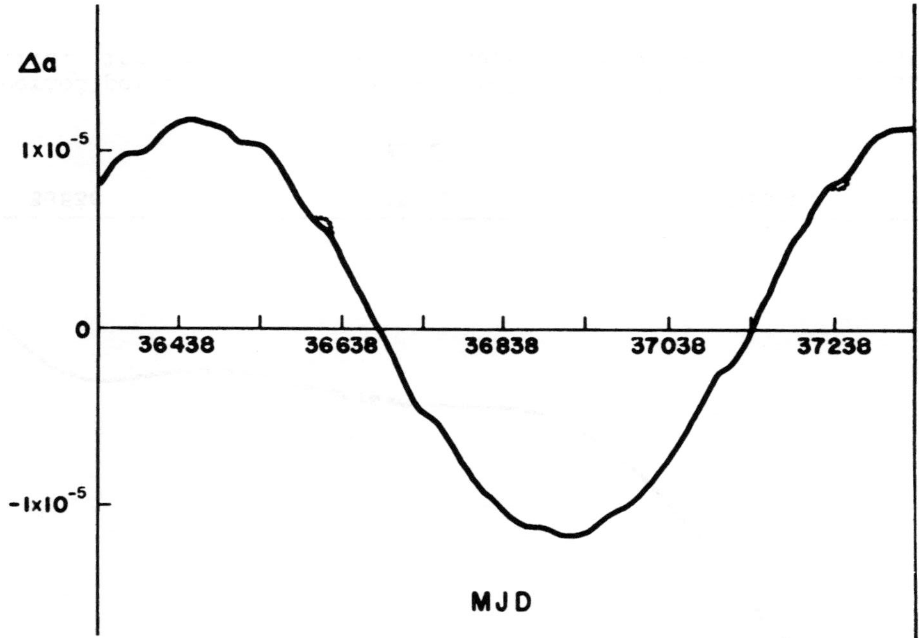

Figure 2. Long period perturbations of the semimajor axis of the Vanguard II satellite. The dotted line results of numerical integration.

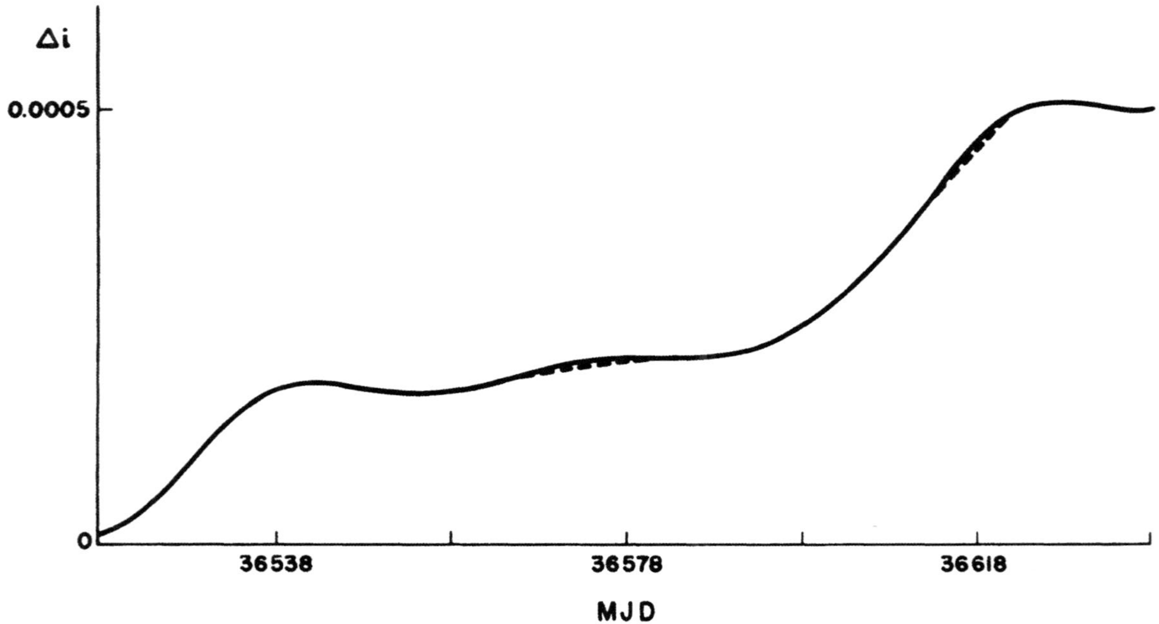

Figure 3. Long period perturbations of the orbital inclination of the Vanguard II satellite. The dashed line shows results obtained by Kozai. (continued.)

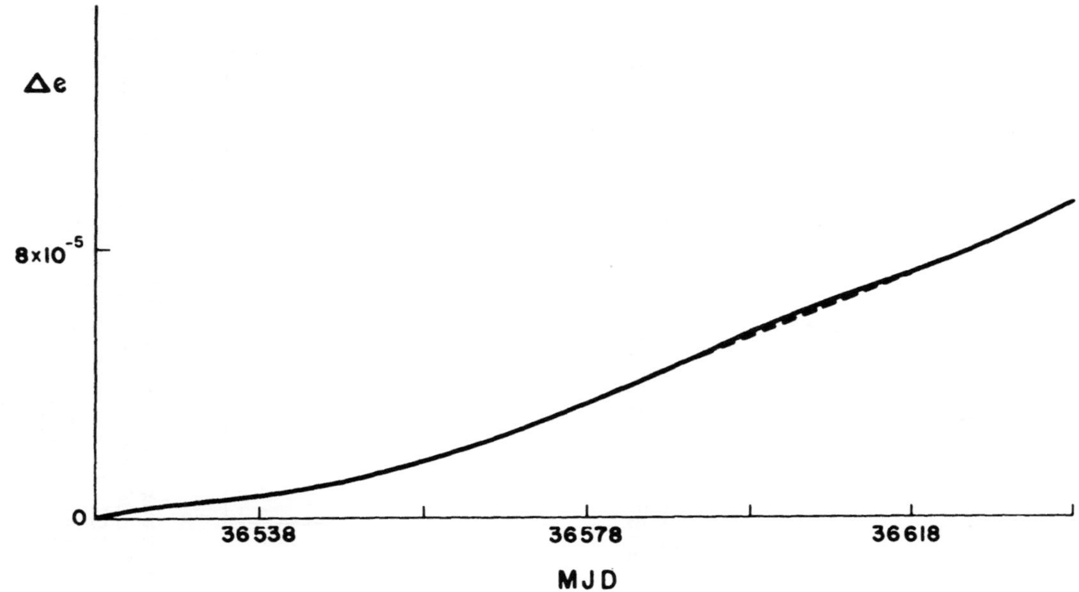

Figure 3. Long period perturbations of the orbital eccentricity of the Vanguard II satellite. The dashed line shows results obtained by Kozai.

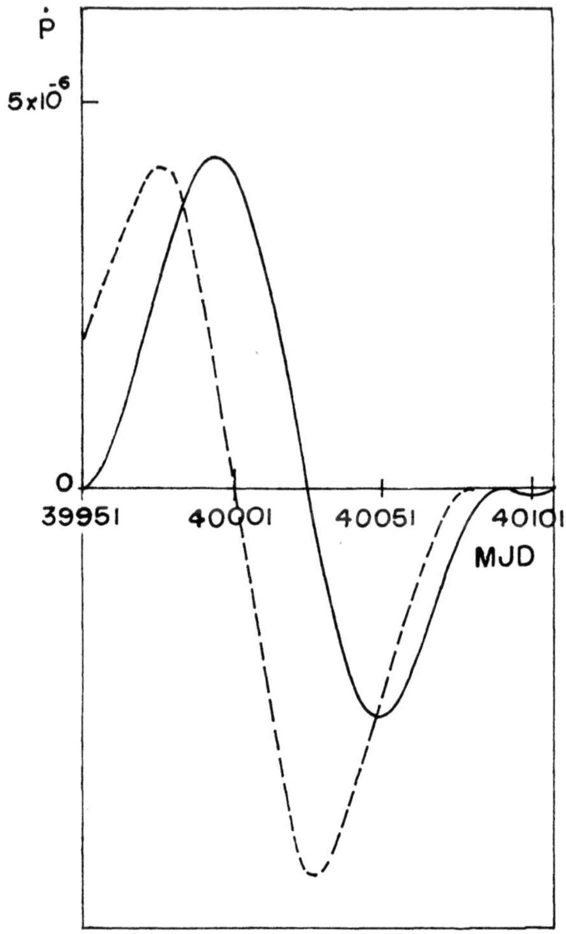

Figure 4. Perturbations of the anomalistic period of the PAGEOS satellite during the period 39951-40101 (MJD), as compared with Lāla's (dashed line). The shift is due to a bad choice of initial condition.

References

1 Mello, F.S.: Action de la Pression de Radiation sur le Mouvement d'un Satellite Artificiel de la Terre. Proc. 14th Int. Astronaut. Congress. Paris, 1963, v. IV, PWN, Warsaw (1965) 41-50.

2 Kovalevsky, J.: Les Calculs de Mécanique Céleste sur Ordinateur. Inst. Tech. Aeronaut. Obs. Astron. São José dos Campos (1971).

3 Kozai, Y.: Effects of Solar Radiation Pressure on the Motion of an Artificial Satellite. Smithsonian Astrophys. Obs. Spec. Rep. 56 (1961) 25-33.

4 Lála, P.: Semi-Analytical Theory of Solar Pressure Perturbations of Satellite Orbits During Short Time Intervals. Bull. Astr. Inst. Cal. 22 (1971) 63-72.

5 Brouwer, D.; Hori, G.I.: Theoretical Evaluation of Atmospheric Drag Effects in the Motion of an Artificial Satellite. Astron. J. 66 (1961) 193-225.

Optimal Impulsive Interplanetary Transfers

JEAN-PAUL GRAVIER[1], CHRISTIAN MARCHAL[2], ROBERT DUDLEY CULP[3]
[1]Visiting Professor, Department of Electrical Engineering, Pontifica Universidade Católica do Rio de Janeiro, Brazil.
[2]Research Scientist, L'Office d'Etudes et de Recherches Aérospatiales, Chatillon, France.
[3]Assistant Professor, Department of Aerospace Engineering Sciences, University of Colorado, Boulder, Colorado 80302

Abstract

This study presents the derivation of the equations necessary to establish a practical method for computing optimal transfers in the real case. It makes it possible to compute and use actual optimal transfers for advanced planning of preliminary mission analysis in place of the standard Hohmann transfers. This realizable solution includes the effects of the inclination and eccentricity of the orbits of the planets. It applies directly for cases in which a two-impulse transfer is the optimal solution; the results are presented as functions of the corresponding idealized Hohmann transfers.

Introduction

One of the major problems of the space age is concerned with changing orbits in space. The stringent requirements usually associated with orbit changing maneuvers make it essential that fuel optimal transfers be investigated. For the existing systems of rocket propulsion the criterion of optimization is the classical characteristic velocity, the arithmetic sum of all the artificial velocity changes. In the real cases of transfer we can no longer admit the result of Hohmann (1), as an economical solution when the terminal orbits are circular and coplanar. As would be expected from the results of the coplanar orbit transfer problem (2,3), the effect of the eccentricities of the planetary orbits is negligible.

But a recent analytical study by Marchal (4) limited to the first order shows that the effect of the inclinations of the orbits of any of the planets is very great. A similar approach is used in this work in order to derive the general equations leading to a procedure of computation applicable to real cases (5). This three dimensional approach considers the problem of transferring a rocket with minimum characteristic velocity from an initial planet on an elliptical orbit to a final planet on a non-coplanar elliptical orbit about the same center of force. In this way optimal trajectories are provided for preliminary mission analysis (6,7). They solve once and for all any particular case whose isolated computation is otherwise lengthy.

The procedure derived is valid for computing the exact optimal transfers between two arbitrary planets, including those cases for which midcourse impulses are required. But numerical applications in this ultimate case are only in the process of being completed.

The originality of the present study is to establish a simple correspondence between the optimal transfer and the idealized Hohmann transfer through a defining parameter called the Hohmann planetary window.

Restrictions and Assumptions Involved in the Transfer

We split the trajectory into three phases so that each phase is in a two-body orbit. To do so we define a mean sphere of influence (MSI) whose boundary is such that the effects of the gravitational attractions between the planets and the sun are nearly equal. For the Earth-Sun system the radius of such a sphere is taken as 10^6 km. The trajectory consists of: a) a planetocentric phase in the vicinity of the initial planet divided into a part through the atmosphere, in which the aerodynamical shape of the vehicle is very important, an extra-atmospheric part of the trajectory and heliocentric phase in interplanetary space; b) an identical planetocentric phase near the final planet.

Our unique concern will be, from now on, to derive the elements of the suitable optimal transfer for the heliocentric phase (Fig. 1). Thus we need only to consider a modified heliocentric problem between parabolic levels.

We assume further that there exists no intermediate fly-by in the MSI of a third planet. Hence we do not have to consider hyperbolic phases. Furthermore we assume that impulses are available: this is not too restrictive because the time of transit required for an interplanetary transfer is very large compared with the thrust phase (hundreds of days compared with a few minutes). In the case of such an intermediate thrust phase of duration τ along the optimal orbit in a Newtonian force field at a distance r from the center of attraction, the relative loss is always smaller than

$$\frac{\mu\tau^2}{24r^2} = \frac{\Pi^2\tau^2}{6T^2}$$

where T is the period of a circular orbit at the altitude r.

Example: At an altitude of 100 km the period of a satellite in circular orbit around the Earth is greater than 86 min, thus for a thrust phase of 2 min we obtain

$$\frac{\Pi^2\tau^2}{6T^2} < 9 \times 10^{-4} \quad .$$

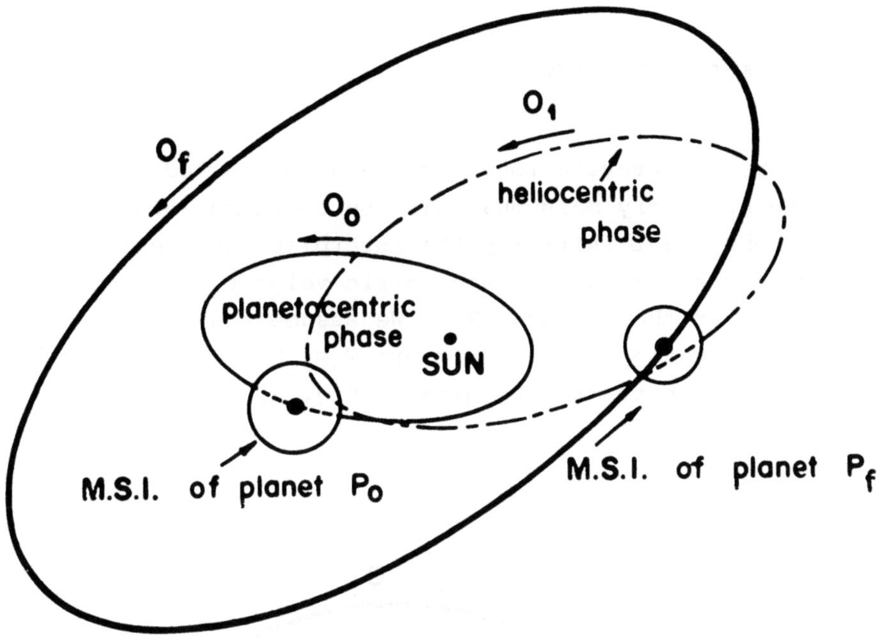

Figure 1. Segmented trajectory in three principal phases.

Modified Velocity Requirements of the Heliocentric Problem

To make the problem purely heliocentric we need to modify the magnitude of the velocity increments to account for the planets. The heliocentric orbit has a very definite inertial velocity which we will call $\underline{V}_{01}$. The departure planet has a specific velocity vector $\underline{V}_{00}$. It follows that $\underline{V}_{R0} = \underline{V}_{01} - \underline{V}_{00}$ is the required velocity vector which must be added to the velocity vector of the escape hyperbola of a rocket to place it on the heliocentric orbit. Hence $\underline{V}_{R0}$ is the planetocentric velocity at the boundary of the MSI. In fact if the planet had no mass, $\underline{V}_{R0}$ would be exactly the additive velocity increment that would have to be provided by the rocket propulsive system.

However, the gravitational acceleration caused by the finite planetary mass must also be overcome. Let it be assumed that initially the interplanetary vehicle has an absolute initial planetocentric velocity L_0 (escape velocity at the distance R). We assume that for a specified trip the heliocentric arc M_0M_f has been determined between the two terminals (Fig. 2). At distance

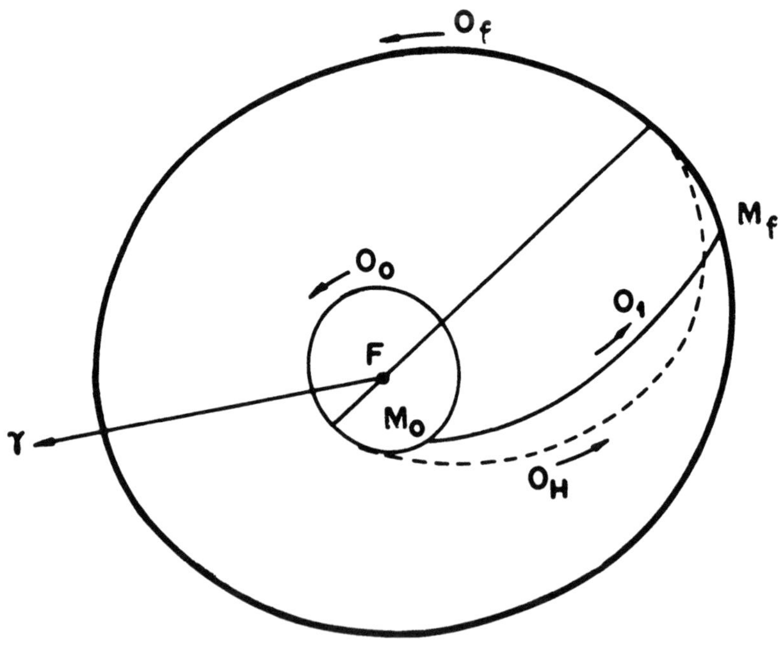

Figure 2. The general problem showing the idealized Hohmann transfer O_h and the corresponding actual optimal transfer O_1 all projected into one plane.

R, the boundary of the atmosphere, taken as 100 km, the rocket with the planetocentric velocity L_0 is given an impulse $\underline{\Delta V}$ which puts the vehicle on the hyperbolic escape arc PA. Because of the long distance M_0M_f of an interplanetary mission (millions of km) compared with the very short orbital segment PA, we may assume that the velocity vector on the arc M_0M_f at A is identical with the velocity vector at M_0 and thus can compute $\underline{V}_{R0} = \underline{V}_{01} - \underline{V}_{00}$ at point A. The energy relation on a hyperbola is:

$$V_R^2 = \mu\left[\frac{2}{r} + \frac{1}{a_h}\right]$$

where V_R is the planetocentric speed, μ is the sum of the masses of the rocket and the planet, a_h is the semimajor axis of the hyperbola. We apply the above relation at point A, on the MSI (10^6km), considered as a very distant point of the atmosphere (10^2km):

$$\lim_{r\to\infty} V_{R0}^2 = \frac{\mu}{a_h} \quad .$$

At point P, to put the rocket on the orbit with velocity V_R at infinity, a velocity increment ΔV must be added to the parabolic level corresponding to velocity L_0:

$$\Delta V = V - L_0 \quad .$$

We apply again the energy relation at point P:

$$V^2 = \mu\left[\frac{2}{R} + \frac{1}{a_h}\right] \quad .$$

Consequently

$$\Delta V = \sqrt{\mu\left[\frac{2}{R} + \frac{1}{a_h}\right]} - L_0 \quad .$$

Making the appropriate substitutions yields

$$\Delta V = \sqrt{L_0^2 + V_{R0}^2} - L_0 \quad .$$

This modification replacing V_R by $\Delta V = \sqrt{V_R^2 + L^2} - L$, allows us to handle the problem as a purely heliocentric one.

Formulation of the Problem

Within the established set of constraints we want to find the kinematic characteristics of the most economical transfer from the initial planet to the final. The minimum characteristic velocity of the hyperbolic escape or capture maneuver at either of the planets is:

$$C = \sqrt{V_R^2 + L^2} - L \quad .$$

When atmospheric drag is taken into account the characteristic velocity of the capture maneuver is zero. Hence between parabolic planetary levels the characteristic velocity of a general optimal maneuver is

$$C = \sqrt{V_{R0}^2 + L_0^2} - L + \Sigma I_i + \sqrt{V_{R_f}^2 + L_f^2} - L_f \quad (1)$$

where the middle term is the characteristic velocity along the open arc of the trajectory.

Optimal Control Problem

The study of this problem of optimization of the heliocentric phase is based on the use of the primary vector (8). The theoretical problem of optimal interplanetary transfer with planets has been reduced to a time-free two-point boundary value problem between the orbits themselves.

The state of the rocket is defined by:

$\underline{r} = \underline{r}(t)$ = radius vector at time t

$\underline{V} = \underline{V}(t)$ = velocity vector at time t

and we want to minimize the function:

$$C = \int_0^t \| \underline{\Gamma}(t) \| \, dt$$

where

$$\frac{dC}{dt} = \| \underline{\Gamma} \| = \Gamma \quad .$$

C, the characteristic velocity, represents a statement of the fuel consumption since departure:

$$\underline{\Gamma} = \frac{T}{m} = \frac{\text{Thrust at time t}}{\text{Rocket mass at time t}} \quad .$$

The differential equations governing the motion of the center of mass are written in the following form:

$$\underline{V} = \frac{d\underline{r}}{dt} \tag{2}$$

$$\frac{d\underline{V}}{dt} = \underline{\Gamma} - \mu \frac{\underline{r}}{r^3} \quad . \tag{3}$$

If we denote by $\underline{s}$ a unit vector in the direction of the thrust then $1 - \underline{s}\cdot\underline{s} = 0$. The boundary conditions to be satisfied are: Initial values: $\underline{r} = \underline{r}_0$; $\underline{V} = \underline{V}_0$ at time t_0. Final values: $\underline{r} = r_f$; $\underline{V} = \underline{V}_f$ at time t_f.

We represent the final conditions by the function by $\psi(\underline{r}_f, \underline{V}_f, t_f)$. The characteristic function of the rocket propulsion system is chosen so that $0 \leq \Gamma(t) \leq \Gamma_{max}$ which gives a constraint on the control Γ. Following the classical technique of optimal control we write the above inequality constraint in the form:

$$\Gamma(t)\left[\Gamma_{max} - \Gamma(t)\right] - \theta^2 = 0$$

where Γ and θ serve as control functions. We introduce two types of Lagrange multipliers, to solve the present Mayer-Bolza problem.

1. The primary vector $\underline{P}_V$, the vector $\underline{P}_r$ and the scalar P_c are used to construct the Hamiltonian:

$$H_p = \underline{P}_r \cdot \frac{d\underline{r}}{dt} + \underline{P}_V \cdot \frac{d\underline{V}}{dt} + P_c \frac{dC}{dt} \quad .$$

Using the equations of motion (2) and (3) this reduces to

$$H_p = \underline{P}_r\cdot\underline{V} - \underline{P}_V \cdot \frac{\mu}{r^3}\,\underline{r} + \underline{P}_V\cdot\underline{\Gamma} + P_c\cdot\Gamma \quad . \tag{4}$$

2. The scalar multipliers μ_1 and μ_2 corresponding to the finite constraints imposed on the control function:

$$H_\mu = \mu_1(1 - \underline{s}\cdot\underline{s}) + \mu_2\left(\Gamma(\Gamma_{max} - \Gamma) - \theta^2\right) \quad .$$

We obtain, in the state space representation, the functional:

$$\underline{x} = \begin{vmatrix} \underline{r} \\ \cdots \\ \underline{V} \end{vmatrix}_{6,1} \qquad \lambda(t) = \begin{vmatrix} P_r \\ \cdots \\ P_V \end{vmatrix}_{6,1} \qquad \underline{\mu}(t) = \begin{vmatrix} \mu_1 \\ \\ \mu_2 \end{vmatrix}_{2,1}$$

so that the state space equation is

$$\dot{\underline{X}} = f(X,\underline{\Gamma},t) = A\underline{X} + B\underline{\Gamma} \tag{5}$$

where

$$A = \left[\begin{array}{ccc|ccc} 1 & 0 & 0 & & & \\ 0 & 1 & 0 & & 0 & \\ 0 & 0 & 1 & & & \\ \hline & & & -\frac{\mu}{r^3} & 0 & 0 \\ & 0 & & 0 & \frac{\mu}{r^3} & 0 \\ & & & 0 & 0 & -\frac{\mu}{r^3} \end{array}\right] \qquad B = \begin{bmatrix} 0 \\ 0 \\ 0 \\ 1 \\ 1 \\ 1 \end{bmatrix}$$

with the functional:

$$J = \psi(\underline{r}_f,\underline{V}_f,t_f) + \int_0^{t_f} \left\{\underline{P}_r \cdot \frac{d\underline{r}}{dt} + \underline{P}_V \cdot \frac{d\underline{V}}{dt} - H_p - H\mu + P_c\Gamma\right\} dt \quad . \tag{6}$$

We then set the variation of the above functional equal to zero to derive the optimal conditions. There are three groups of relations: the equations of stationary state, the Weierstrass-Erdmann corner conditions at time t^* of possible lack of continuity of the controls, and the boundary conditions that the Lagrange multipliers $\underline{P}_r$, $\underline{P}_V$, P_c and the Hamiltonian must satisfy.

We form the function E, corresponding to the rocket propulsion system:

$$E = \underline{P}_V \cdot (\underline{\Gamma} - \underline{\Gamma}^*) + P_c(\Gamma - \Gamma^*) \quad . \tag{7}$$

This function must be such that $E > 0$, to satisfy the fundamental Weierstrass necessary criterion for a minimum. The asterisk denotes an admissible control which satisfies the constraints imposed on the controls. The optimal requirements give consequently the set of relations:

$$\frac{d\underline{P}_v}{dt} = -\frac{\partial H_p}{\partial \underline{v}} = -\underline{P}_r \tag{8}$$

$$\frac{dP_c}{dt} = -\frac{\partial H_p}{\partial \underline{r}} = \frac{\mu}{r^3}\left[\underline{P}_v - 3\,\frac{\underline{P}_v \cdot r}{r^3}\,\underline{r}\right] \tag{9}$$

$$\frac{dP_c}{dt} = \frac{\partial H_p}{\partial c} = -\,(P_v+P_c)\,\frac{\partial \Gamma}{\partial c} = -\,\frac{1}{\Gamma}\,(P_v+P_c)\,\frac{\partial^2 C}{\partial t^2} \tag{10}$$

$$\frac{\partial (H_p+H_\mu)}{\partial \underline{s}} = \Gamma\underline{P}_v - 2\mu_1\underline{s} = 0 \tag{11}$$

$$\frac{\partial (H_p+H_\mu)}{\partial c} = (P_v+P_c)\,\frac{\partial \Gamma}{\partial c} + \mu_2(\Gamma_{max}-2\Gamma)\,\frac{\partial \Gamma}{\partial c} = 0 \tag{12}$$

$$\frac{\partial H_\mu}{\partial \theta} = -\,2\mu_2\theta = 0 \quad . \tag{13}$$

From Eq. (11) it follows that the primary vector is parallel to the thrust. From the Weierstrass condition we also deduce that

$$\underline{P}_v = P_v\underline{s} \quad . \tag{14}$$

Upon eliminating $\underline{P}_r$ from Eqs. (8) and (9) we derive a second-order differential equation for the primary vector $\underline{P}_v$:

$$\frac{d^2\underline{P}_V}{dt^2} = -\frac{\mu}{r^3}\left\{\underline{P}_V - 3\,\frac{\underline{P}_V\cdot r}{r^2}\,\underline{r}\right\} \quad . \qquad (15)$$

We introduce $\eta = P_V + P_C$, and Eq. (12) becomes

$$\eta + \mu_2(\Gamma_{max} - 2\Gamma) = 0 \qquad (16)$$

and the expression for the Hamiltonian

$$H_p = \frac{d\underline{P}_V}{dt}\cdot\underline{V} - \frac{\mu}{r^3}\,\underline{P}_V\cdot\underline{r} + \eta\Gamma \quad . \qquad (17)$$

To solve Eq. (13), the assumption $\theta = 0$ gives $\Gamma(t) \times (\Gamma_{max} - \Gamma(t)) = 0$. Consequently H_p is maximized if $\Gamma = 0$, when $\eta < 0$, corresponding to the motion along a Keplerian arc. $\Gamma = \Gamma_{max}$ when $\eta > 0$, the regime of maximum thrust.

At the instant of switching, the Lagrange multipliers and consequently η suffer no discontinuity. Thus we obtain: $\eta(t^*) = 0$.

The other solution of Eq. (13) is $\mu_2 = 0$ which corresponds to an arc of "singular control" with variable thrust when $0 < \Gamma(t) < \Gamma_{max}$.

We refer to Marchal (9) and Kelley (10) for an extensive discussion of the arcs of intermediate continuous thrust. Thus the impulsive thrust studied here is a limiting case when the maximum thrust Γ_{max} tends to infinity. At a junction point of switching between a phase with or without thrust: $\eta(t^*) = 0$. From Eq. (10)

$$\lim_{\Gamma_{max}\to\infty} \frac{dP_c}{dt} = - \frac{\eta}{\Gamma_{max}} \frac{\partial^2 C}{\partial t^2} = 0 \quad . \tag{18}$$

Thus P_c has a fixed value $P_c = 1$ and consequently, at the impulses, $\| \underline{P}_V \| = 1$; therefore

$$P_V = \frac{V_R}{\sqrt{V_R^2 + L^2}}$$

at the boundary of the MSI (see section on Transfer between Parabolic Planetocentric Levels).

Scalar Integral

Since for an impulse P_c is constant and Γ is zero along ballistic arc, equations of motion together with Eq. (15) give

$$\frac{dH_p}{dt} = \frac{\mu}{r^3}\left[\underline{P}_V - 3\left[\frac{\underline{P}_V \cdot \underline{r}}{r^3}\right] r\right] \cdot \underline{V} - \frac{d\underline{P}_V}{dt}\left[-\mu \frac{\underline{r}}{r^3} + \underline{\Gamma} + \mu \frac{\underline{r}}{r^3}\right]$$

$$+ 3\mu \frac{\underline{P}_V \cdot r}{r^3} \left(\frac{\underline{r} \cdot \underline{V}}{r^2}\right) - \frac{\underline{P}_V \cdot \underline{V}}{r^3} + \frac{dP_V}{dt} = 0 \quad . \tag{19}$$

This proves the constancy of the Hamiltonian along an optimal trajectory:

$$H_p = \text{constant} = h_p \quad . \tag{20}$$

Vector Integral

We perform the vector product of both sides of Eq. (15) with the vector $\underline{r}$ which leads to

$$\frac{d}{dt}\left[\frac{d\underline{P}_V}{dt} \times \underline{r} - \underline{P}_V \times \underline{V}\right] = 0 \quad . \tag{21}$$

After integration we obtain

$$- \frac{d\underline{P}_V}{dt} \times \underline{r} + \underline{P}_V \times \underline{V} = \underline{k} \quad . \tag{22}$$

Consequently the two first order differential Eqs. (20) and (22) are equivalent to Eq. (15). A new integration gives the analytical expression of the primary vector:

$$\underline{P}_V = \Lambda_1 V + \Lambda_2 \times \underline{r} + \Lambda_3 (2\underline{r} - 3\underline{V}t) + (r \times \underline{V}) \times \underline{\Lambda}_4$$
$$+ \underline{r} \times (\underline{V} \times \underline{\Lambda}_4) \tag{23}$$

which depends upon eight arbitrary constants Λ_1, $\underline{\Lambda}_2$, Λ_3, $\underline{\Lambda}_4$.

The following two relations for trajectories in a central gravitational field hold true:

$$\underline{e} \times \underline{r} + \underline{r} \times \left\{ \underline{V} \quad \frac{\underline{h}}{\mu} \right\} = 0$$

$$-P\underline{V} + \frac{\underline{h}}{a} \times \underline{r} + \underline{h} \times \underline{e} + \underline{r} \times (\underline{V} \times \underline{e}) = \underline{0} \tag{25}$$

where $\underline{e}$ is the eccentricity vector and $\underline{h}$ is the vector angular momentum.

These relations allow us to reduce the set of arbitrary constants to six, because for

$$\Lambda_1 = 0 \quad \underline{\Lambda}_2 = \underline{e} \quad \Lambda_3 = 0 \quad \underline{\Lambda}_4 = \frac{\underline{h}}{\mu}$$

the relation (Eq. 24) gives identically

$$\underline{P}_{V1} \left\{ 0, \underline{e}, 0, \frac{\underline{h}}{\mu} \right\} \equiv \underline{0} \tag{26}$$

and for

$$\Lambda_1 = -P \quad \underline{\Lambda}_2 = \frac{\underline{h}}{a} \quad \Lambda_3 = 0 \quad \underline{\Lambda}_4 = \underline{e}$$

the relation (25) gives identically

$$\underline{P}_{v2}\left(-P, \frac{\underline{h}}{a}, 0, \underline{e}\right) \equiv \underline{0} \quad .$$

Consequently if we write the following linear combination in the classical S, T, W coordinate system (radial and horizontal components in the osculating plane of motion, and the component normal of these):

$$\underline{P}_v = \underline{P}_v - \frac{\Lambda_{4z}}{h} \underline{P}_{v1} - \frac{\Lambda_{4x}}{e} \underline{P}_{v2}$$

$$\underline{P}_v = \lambda_1 V + \underline{\lambda}_2 \times \underline{r} + \lambda_3 (2\underline{r} - 3Vt) + \underline{h} \times \lambda_4 + \underline{r} \times (\underline{V} \times \underline{\lambda}_4) \tag{28}$$

The new constants are

$$\lambda_1 = \Lambda_1 + \frac{\Lambda_{4x}}{e} P$$

$$\lambda_2 = \begin{bmatrix} \lambda_{2x} = \Lambda_{2x} - \mu \dfrac{\Lambda_{4x}}{h} e \\ \lambda_{2y} = \Lambda_{2y} \\ \lambda_{2z} = \Lambda_{2z} - \dfrac{\Lambda_{4x} h}{e \cdot a} \end{bmatrix}$$

$$\lambda_3 = \Lambda_3$$

$$\lambda_4 = \begin{bmatrix} \lambda_{4x} = 0 \\ \lambda_{4y} = \Lambda_{4y} \\ \lambda_{4z} = 0 \end{bmatrix} \quad .$$

Constant Hamiltonian

To find the actual value of the constant h_p of the

Hamiltonian we make use of the scalar integral and expression (28) together with the energy relation along an ellipse

$$V^2 - 2\frac{\mu}{r} = -\frac{\mu}{a_1} \quad .$$

Thus

$$h_p = -\frac{\lambda_3 \mu}{a_1} \quad . \qquad (29)$$

Constant Vector $\underline{k}$

The vector integral relation (22) expressed in the (S; XYZ) coordinate system gives for the vector $\underline{k}$:

$$k_x = h\lambda_{2y}$$

$$k_y = -h\lambda_{2x}$$

$$k_z = h\left(\lambda_3 - e_1 \frac{h}{P_1} \lambda_{4y}\right) \quad . \qquad (30)$$

Primary Vector $\underline{P}_v$

We need to express the primary vector in the coordinate system S, T, W moving along the intermediate orbit. So we perform a rotation of angle v in the plane of the orbit:

$$\begin{bmatrix} P_{v_S} \\ P_{v_T} \\ P_{v_W} \end{bmatrix} = \begin{bmatrix} \cos v & \sin v & 0 \\ -\sin v & \cos v & 0 \\ 0 & 0 & 1 \end{bmatrix} \quad .$$

Hence

$$P_{v_S} = \frac{\lambda_1 h}{P} e \sin v - h\lambda_{4y}\cos v + \lambda_3 a(1 - e\cos E)$$

$$- \frac{3e \sin v}{(1-e^2)^{1/2}} (E - e \sin E)$$

$$P_{v_T} = \frac{\lambda_1 h}{P} (e + \cos v) + h\lambda_{4y} \sin v + \lambda_{2z} r + h\lambda_{4y}$$

$$\times \frac{\sin E}{(1-e^2)^{1/2}} + \lambda_3 \frac{-3a}{1-e^2} (1 + e \cos v)(E - e \sin E)$$

$$P_{v_W} = a(\lambda_{2x}(1-e^2))^{1/2} \sin E - \lambda_{2y}(\cos E - e) \quad .$$

Next we find the classical relations

$$P_{v_S} = \nu_3 e \sin v - \nu_5 \cos v$$

$$+ \nu_6 \left[2(1-e \cos E) - \frac{3e \sin v}{(1-e^2)^{1/2}} (E - e \sin E)\right]$$

$$P_{v_T} = \nu_3 (1 + e \cos v) + \nu_4 (1 - e \cos E) + \nu_5 \left\{\sin v + \right.$$

$$\left. + \frac{\sin E}{(1-e^2)^{1/2}}\right\} - \nu_6 \frac{3(1 + e \cos v)}{(1-e^2)^{1/2}} (E - e \sin E)$$

$$P_{v_W} = \nu_1 \sin E + \nu_2 (\cos E - e) \tag{31}$$

where the following relations between the λ's and the ν's coefficients hold true:

$$\nu_1 = a\lambda_{2x}(1-e^2)^{1/2} \qquad \nu_4 = a\lambda_{2z}$$

$$\nu_2 = -a\lambda_{2y} \qquad \nu_5 = h\lambda_{4y}$$

$$\nu_3 = (\mu/p)^{1/2} \qquad \nu_6 = a\lambda_3 \quad . \tag{32}$$

For this gives

$$H_p = \frac{\mu}{a_1^2} \nu_6 = -n_1^2 a_1 \nu_6 \tag{33}$$

$$\underline{k} = \frac{h}{a_1}\begin{bmatrix} -\nu_2 \\ \dfrac{\nu_1}{(1-e^2)^{1/2}} \\ \nu_6 - \dfrac{e}{1-e_1^2}\nu_5 \end{bmatrix} \quad . \tag{34}$$

Time Derivative of the Primary Vector

We obtain easily:

$$\frac{d\underline{P}_V}{dt} = \lambda_1\mu \frac{\underline{r}}{r^3} + \lambda_2 \times \underline{V} + \lambda_3(-\underline{V} + 3\frac{\mu t}{r^3}\underline{r})$$

$$+ \underline{V}\times(\underline{V}\times\lambda_4) - \frac{\mu}{r^3}\underline{r}\times(\underline{r}\times\lambda_4) \quad . \tag{35}$$

In the rotating system (S,T,W) the components of this vector are

$$\left(\frac{dP_V}{dt}\right)_S = -\nu_3 \frac{\mu^2}{h^3}(1 + e\cos v)^2 - \nu_4 \frac{\mu}{ha}(1 + e\cos v)$$

$$- \nu_5 \frac{\mu^2}{3}\sin v(1 + e\cos v)$$

$$+ \nu_6\left[-\frac{\mu e}{ha}\sin v + \frac{3\mu^2 M(1 + e\cos v)^2}{h^3(1-e^2)^{1/2}}\right]$$

$$\left(\frac{dP_V}{dt}\right)_T = \nu_4 \frac{\mu e}{ha}\sin v + \nu_5 \frac{\mu^2}{h^3}(e + \cos v)$$

$$- \nu_6 \frac{\mu}{ha}(1 + e\cos v)$$

$$\left(\frac{dP_v}{dt}\right)_W = \nu_1 \frac{\mu(e + \cos v)}{ha(1-e^2)^{1/2}} - \nu_2 \frac{\mu \sin v}{ha} \qquad (36)$$

where h is the angular momentum, the semi-major axis is a and μ is the solar gravitational constant.

Transfer between Parabolic Planetocentric Levels

In this section we derive the end conditions for the primary vector as outlined in the section on Optimal Control Problem. Using the approximation of the MSI, we can neglect the field of the sun whilst the rocket is in the hyperbolic escape orbit. Along this trajectory we obtain the components of the primary vector (31) where e_h and v are respectively the eccentricity and the true anomaly on the escape hyperbola. The six constants of integration are now called $\mu_1 \ldots \mu_6$. An impulsive thrust is applied at the apse of the hyperbola in the direction of motion. At this point the thrust is parallel to the primary vector and the condition of optimality $\underline{P}_v(d\underline{P}_v/dt) = 0$ must be satisfied.

The conditions together with v = 0 at the apse give

$$P_{v_S} = -\mu_5 + 2\mu_6(1-e_h) = 0$$

$$P_{v_T} = \mu_3(1+e_h) + \mu_4(1-e_h) = 1$$

$$P_{v_W} = \mu_2(1-e_h) = 0$$

$$\dot{P}_{v_S} = \left(\frac{dP_v}{dt}\right)_S + P_{v_T}\left(\frac{dP_v}{dt}\right)_T + P_{v_W}\left(\frac{dP_v}{dt}\right)_W$$

$$= \left(\frac{\mu}{P_h}\right)^{1/2} \left(\frac{\mu_5}{P_h} - \frac{\mu_6}{a_h}\right)(1+e_h) \quad ;$$

it follows that

$$\mu_2 = \mu_5 = \mu_6 = 0 \quad \mu_4 = \frac{1}{1-e_h}\,(1 - (1+e_h)\mu_3)$$

and therefore the components of the primary vector reduce to

$$P_{v_S} = e_h\mu_3 \sin v$$

$$P_{v_T} = \mu_3(1+e_h \cos v) + \frac{1+e_h - (1+e_h)^2\mu_3}{1+e_h \cos v}$$

$$P_{v_W} = \mu_1(e_h^2-1)^{1/2}\,\frac{\sin v}{1+e_h \cos v} \quad .$$

The components of the derivative are

$$\left(\frac{p_h}{\mu}\right)^{1/2}\left(\frac{dP_v}{dt}\right)_S = \mu_3\,\frac{(1+e_h \cos v)^2}{p_h} - (1 - (1+e_h))\mu_3\,\frac{1+e_h \cos v}{a_h(1-e_h)}$$

$$\left(\frac{p_h}{\mu}\right)^{1/2}\left(\frac{dP_v}{dt}\right)_T = \frac{e_h \sin v}{a_h(1-e_h)}\,(1 - (1+e_h))\mu_3$$

$$\left(\frac{p_h}{\mu}\right)^{1/2}\left(\frac{dP_v}{dt}\right)_W = \mu_1\,\frac{e_h + \cos v}{a_h(e_h^2-1)^{1/2}} \quad .$$

The polar equation of the hyperbolic escape trajectory is

$$r_h = \frac{p_h}{1+e_h \cos v} = \frac{a_h(e_h^2-1)}{1+e_h \cos v} \quad .$$

At a very remote point on the trajectory the limiting values are when $r = \infty$:

$$\cos v = -\frac{1}{e_h} \qquad \sin v = \frac{(e_h^2-1)^{1/2}}{e_h} \quad .$$

We consider that the point A (section on Time Derivative of the Primary Vector) of entry in interplanetary transfer orbit satisfies this hypothesis and this gives for the primary vector and its derivative:

$$P_{v_S} = \mu_3 (e_h^2 - 1)^{1/2}$$

$$P_{v_T} = \frac{r_h}{p_h} (1+e_h)(1 - (1+e_h)\mu_3)$$

$$P_{v_W} = \frac{r_h}{p_h} (e_h^2-1)\mu_1 \qquad (37)$$

$$\left(\frac{dP_v}{dt}\right)_S = 0$$

$$\left(\frac{dP_v}{dt}\right)_T = \frac{\mu^{1/2}}{p_h^{3/2}} (e_h^2-1)^{1/2} (1+e_h)\left|1 - (1+e_h)\mu_3\right|$$

$$\left(\frac{dP_v}{dt}\right)_W = \frac{\mu^{1/2}}{p_h^{3/2}} \frac{(e_h^2-1)^{3/2}}{e_h} \mu_1 \quad . \qquad (38)$$

Upon capture or escape at either of the planets we must express the fact that the components of the primary vector along the interplanetary orbit are identical with those along the hyperbolic orbit. Since the instants of departure and arrival are not specified the expression for $\underline{P}_v$ along the elliptic orbit reduces to

$$\underline{P}_v = \lambda_1 \underline{V} + \underline{\lambda}_2 \times \underline{r} + (\underline{r}\times\underline{V}) \times \underline{\lambda}_4 + \underline{r} \times (\underline{V}\times\underline{\lambda}_4) \qquad (39)$$

with $\lambda_3 = \nu_6/a = 0$ for time independence.

The expression of $\underline{P}_v$ (Eq. 39) in the S,T,W coordinate system leads to the relations

$$\mu_3 = \frac{1}{1+e_h} \qquad \mu_1 = 0 \quad ,$$

and consequently the primary vector (Eq. 37) upon entry into the transfer ellipse is

$$P_{v_S} = \left[\frac{e_h - 1}{e_h + 1}\right]^{1/2} \qquad P_{v_T} = P_{v_W} = 0 \quad . \tag{40}$$

We transform the above expression

$$P_v = \left(\frac{e_h - 1}{e_h + 1}\right)^{1/2} = \frac{1}{\left(1 + \frac{2}{e_h - 1}\right)^{1/2}} \quad .$$

At the periapsis of the hyperbolic trajectory we have $v = 0$. Consequently

$$r_h = R = a_h(e_h - 1) \quad .$$

Hence

$$P_v = \frac{1}{\left(1 + \frac{2a_h}{R}\right)^{1/2}} = \left(\frac{\mu/a_h}{\mu\left(\frac{2}{R} + \frac{1}{a_h}\right)}\right)^{1/2} \quad .$$

Using the relations

$$V_R^2 = \frac{\mu}{a_h} \qquad L^2 = \frac{2\mu}{R}$$

we finally obtain

$$P_v = \frac{V_R}{\sqrt{V_R^2 + L^2}}$$

which justifies the statement of the section on Optimal Control Problem.

Calculation of the Optimal Transfer

The originality of the numerically calculated solution presented here is that it relates the optimal transfer to the well known corresponding Hohmann transfer.

Conditions of Optimality

We look for heliocentric impulsive transfers with:

a) at the departure point M_0, an impulse of amplitude $\sqrt{V_{R_0}^2 + L_0^2} - L_0$ with the associated primary vector

$$\underline{P}_{v_0} = \frac{\underline{V}_{R_0}}{\sqrt{V_{R_0}^2 + L_0^2}}$$

b) $\|\underline{P}_v\| \leq 1$ at all points of the non-thrust phase

c) $\|\underline{P}_v\| = 1$ at all points of eventual intermediate thrust

d) at the arrival point M_f, an impulse of amplitude $\sqrt{V_{R_f}^2 + L_f^2} - L_f$ with associated primary vector:

$$\underline{P}_{v_f} = \frac{-\underline{V}_{R_f}}{\sqrt{V_{R_f}^2 + L_f^2}}$$

or $\underline{P}_{v_f} = 0$ if atmosphere drag is taken into account.

e) the primary vector perpendicular to its time derivative for time-free optimization:

$$\underline{V}_{R_0} \cdot \left(\frac{d\underline{P}_v}{dt}\right)_0 = 0$$

$$\underline{V}_{R_f} \cdot \left(\frac{d\underline{P}_v}{dt}\right)_f = 0 \tag{41}$$

or alternatively for a specified time of transfer $(t_f - t_0)$

$$\underline{V}_{R_0} \cdot \left(\frac{d\underline{P}_v}{dt}\right)_0 = \underline{V}_{R_f} \cdot \left(\frac{d\underline{P}_v}{dt}\right)_f \quad .$$

Technique of Computation

For each possible two-impulse transfer there exists a three-parameter family of orbits. It seems natural to choose the positions of departure and arrival defined by their respective true anomalies as two of these parameters; the third parameter, W, defines the position of the transfer orbit. The exact definition of W is given in Fig. 3. The optimal condition corresponds to the

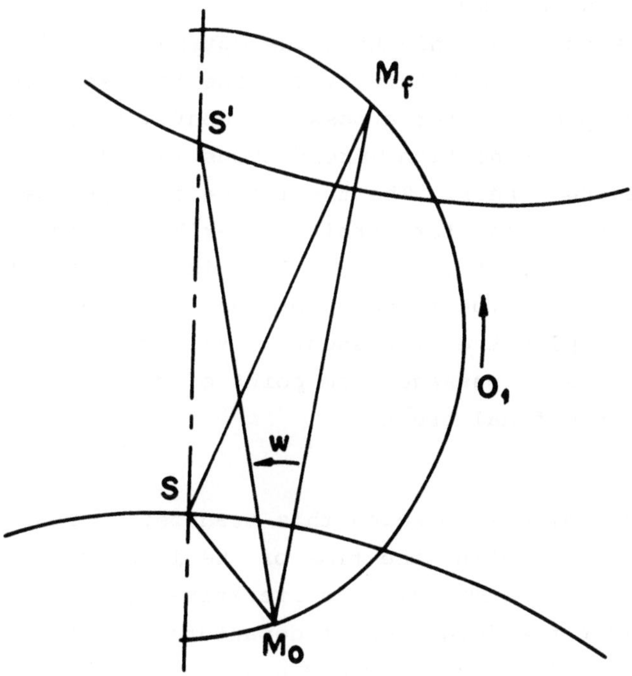

Figure 3. Geometry in the transfer plane showing the hyperbola which is the locus of the second focus of the transfer orbit as a function of w.

minimum of velocity increment. An iterative procedure is developed in (11) to determine this minimum. Briefly it consists in using the first-order results (4) to assess a priori values of the positions of departure from initial orbit, arrival at final orbit and angle W and then compute

$$\underline{V}_{R_0} \cdot \left(\frac{d\underline{P}_v}{dt}\right)_0 = \alpha \quad ; \quad \underline{V}_{R_f} \cdot \left(\frac{d\underline{P}_v}{dt}\right)_f = \beta \qquad (42)$$

with iteration continuing until α and β are as small as desired.

Rendezvous Requirement

Up to this point we only know the optimal transfer joining two planetary orbits. It remains to associate with this time-open transfer a possible configuration of the planets. This condition of rendezvous is calculated in the following manner. The total time from perihelion in the initial planetary orbit to perihelion in the final planetary orbit is found. This is the time of passage from perihelion to point of departure in the initial orbit, plus the time spent in the transfer orbit, plus the time of passage from point of arrival to perihelion in the final orbit.

The Hohmann window to which this transfer corresponds is found by computing the time of the idealized Hohmann transfer in the same way. An iterative process allows us to adjust the longitude of departure for the Hohmann transfer until the times via Hohmann or via optimal transfer are equal. This gives the Hohmann window (defined by the longitude of the Hohmann transfer) to which this transfer belongs.

Results

It is well known (4) that a short and a long transfer correspond to each Hohmann window. The above procedure is repeated for any number of departure points until each of the optimal transfers is expressed as a continuous function of the longitude of the corresponding Hohmann window. Thus, the optimal transfers in the real cases (6,7) are expressed for all possible times as a function over the finite range of 0 to 360 degrees. All the relevant properties of the true optimal maneuver can be simply looked upon as a function of the departure longitude of the Hohmann window.

Dependence of the Characteristic Velocity upon the Planetary Window

The Hamiltonian constant along an optimal trajectory differs of course for each trajectory. We recall the expression of the Hamiltonian (4):

$$H_p = \underline{P}_r \cdot \underline{V} + \underline{P}_v \left(\underline{\Gamma} \cdot \mu \frac{\underline{r}}{r^3} \right) + P_c \frac{dC}{dt}$$

with $P_c = -1$ and

$$\frac{dC}{dt} = f\Gamma \begin{cases} f = 1, \text{ ballistic arc} \\ f = \dfrac{V_R}{V_B} = \dfrac{\partial C}{\partial V_R} \\ \quad \text{in the vicinity of a planet} \end{cases} .$$

Hence

$$H_p = \underline{P}_r \cdot \underline{V} + \underline{P}_v \left(\underline{\Gamma} - \mu \frac{\underline{r}}{r^3} \right) - f\Gamma \quad . \qquad (43)$$

Consequently we obtain

$$dC = \underline{P}_r \cdot \underline{V} dt + \underline{P}_v \left(\underline{\Gamma} - \mu \frac{\underline{r}}{r^3} \right) dt - H_p dt \quad . \qquad (44)$$

The variation of the characteristic velocity between the initial and final time is:

$$dC = (\underline{p}_r \cdot \underline{\delta r} + \underline{p}_V \cdot \underline{\delta V} - H_p \delta t)_0^f \quad ; \tag{45}$$

this gives finally:

$$dC = H_p(dt-dt_0) - (H_p - \underline{p}_{r_f} \cdot \underline{V}_{r_f}) + (H_p - \underline{p}_{r_0} \cdot V_{R_0}) \quad . \tag{46}$$

When we relate this expression to the planetary window we find

$$dC = H_p\left(\frac{1}{n_f} - \frac{1}{n_0}\right) d\omega_H$$

where n is the mean angular motion.

Conclusions

This approach permits the solution of the optimal interplanetary transfer as a simple transfer. The originality of the method is to solve the problem as a simple transfer time-open and to satisfy indirectly the rendezvous requirement. When applied to real planetary cases (6,7), this technique provides all the data necessary for quick orbit calculation for preliminary missions analysis. It should prove to be as easy to use as the idealized Hohmann transfers but is realizable and more accurate.

The computations of transfers in which more than two impulses are required can be considered as an extension of this work and lead to further investigations.

References

1 Hohmann, W.: Die Erreichbarkeit der Himmelskörper. Oldenbourg, München (1925).

2 Marchal, C.: Transferts Optimaux entre Orbites Eliptiques Coplanaires (Durée indifférente). Astro. Acta 2 (1965) 432-445.

3 Culp, R.D.: Coutensou-Busemann Conditions for Optimal Coplanar Orbit Transfer. AIAA J. 5 (1967) 371-372.

4 Marchal, C.: Généralisation des transferts de Hohmann au cas des orbites planétaires faiblement excentriques et faiblement inclinées. ONERA n. 128. Châtillon (1968).

5 Gravier, J.P.; Marchal, C.; Culp, R.D.: A Technique for Obtaining Optimal Interplanetary Transfers in the Real Case. J. Astro. Sci. 18 (1970) 101-117.

6 Gravier, J.P.; Marchal, C.; Culp, R.D.: Optimal Trajectories Between Earth and Mars in their True Planetary Orbits. J. Optimization Techniques and Applications 9 n. 2 (1972).

7 Gravier, J.P.; Marchal, C.; Culp, R.D.: Effects of Inclination and Eccentricity on Optimal Trajectories Between Earth and Venus. Astron. Acta 18 (1973).

8 Lawden, D.F.: Interplanetary Rocket Trajectories. Advances in Space Sciences v. 1. New York and London: Academic Press 1959, 1-53.

9 Marchal, C.: Géneralisation tridimensionnelle et étude de l'optimalité des arcs à poussée intermédiaire de Lawden. La Recherche Aérospatiale n. 123 (1968).

10 Kelley, H.J.; Kopp, R.E.; Gardner Moyer, H.: Singular Extremals. Topics in Optimization. Academic Press 1967.

11 Gravier, J.P.: Absolute Optimum and Two-impulse Optimal Interplanetary Transfers with Applications to Earth-Mars and Earth-Venus. Ph.D. thesis, University of Colorado, 1970.

On the Effect of Earth Radiation Pressure in the Motion of Pageos. Comparisons Between Different Models

D. GAMBIS
G.R.G.S., Meudon Observatory
France

Abstract

The influence of Earth radiation pressure on the motion of an artificial satellite is studied. The case of Pageos has been chosen as an especially significant one. One first points out the effect of the diffuse component and presents a model of it that takes into account the transverse acceleration due to the asymmetry of the illuminated part of the Earth. Then a comparison is made between observation and different models including a) direct radiation pressure and air drag, b) the same plus previous Earth radiation pressure representations, c) the same plus our representation. The advantages and the limitations of our model are put into evidence.

Introduction

The purpose of the present study is to improve the existing model for a better representation of the variations of the orbital elements due to the Earth radiation pressure. A numerical approach is used. In principle, it is always possible to define very sophisticated models but this implies large computer time and this is not really necessary. Therefore the improvement is limited to one point: starting from the model proposed by Wyatt (1), a transverse acceleration is added. Mean orbital elements of Pageos provided by the ephemerides center are used in order to test the theory (Smithsonian Astrophysical Observatory, Goddard Space Flight Center).

Components of the Earth Radiation Pressure

Several components must be considered.

Infra-red component: It seems reasonable to assume that the acceleration produced by this component is radial and constant in a first approximation for the various positions of the satellite (assuming a quasi-circular orbit). Therefore, at this step, this force has been ignored.

Specular component: This component is due to the solar radiation reflected by the Earth. This part is assumed at this stage not to be important in comparison with the diffuse component. The surface of oceans is not quiet and the surface of the ground cannot give a specular contribution. This component has not been considered here.

Diffuse component: This component is due to the solar radiation rediffused by the Earth. It is assumed to be constant and isotropic. The Earth albedo is taken equal to 0.4.

Models of the Acceleration Produced by the Diffuse Component

In the models proposed by Wyatt, the acceleration is assumed to be radial. The shadow effect can be taken into account but there is no transition function. The acceleration becomes instantaneously zero. There exist two versions of the model, called AD1 and AD2.

In our case, the spherical cap of the Earth with the subsatellite point as a pole is divided in small parts (Fig. 1). The effect is obtained by summation of the pressure produced by the different small parts. When the subsatellite point is near to the shadow, the produced acceleration is not radial since the distribution of the illuminated regions is not symmetrical. This is the main difference with the previous models. This model is called REDIF model: there is a transverse force in this case.

Comparison Between Theory and Observations

The variation of the semi-major axis is given in Fig. 2. We restrict our study to the period when the satellite is never in the shadow of the Earth. Then the direct effect of the solar radiation pressure on the semi-major axis is very small. The effects of the solar radiation pressure and of the air drag are computed by a numerical integration (2). An empirical coefficient must be adjusted for the air drag (3). The comparison with the observed variations of the semi-major axis a is made by summing these effects (Fig. 3). Even when multiplying the air drag effect by an empirical number, the shape of the curve is different from the observed shape and no agreement can be obtained. Now, if we add the effect due to the diffuse component (AD1 and AD2 models), the comparison of the shapes shows similarities but not satisfactory agreement (Fig. 4). Finally, if we add the effect computed from the REDIF model, the comparison becomes quite satisfactory (Fig. 5). Such a comparison can be made for many other periods. However, in some cases, significant discrepancies appear (Fig. 6), and it seems impossible to reduce them by using only our theory.

Conclusion

A better agreement has been obtained in many cases, but other effects should be taken into account. For instance, the effect of the latitudinal variation of the Earth albedo is significant (4). Maybe the effect of the infrared and specular component should be introduced. The effect of the gravitational potential of the Earth should also be taken into account with a sufficient accuracy. Finally, the effect of the Earth radiation pressure on the STARLETTE satellite is shown in Fig. 7. STARLETTE is a very dense sphere: the area to mass ratio is equal to 0.1 cgs. More details are available in another study (5).

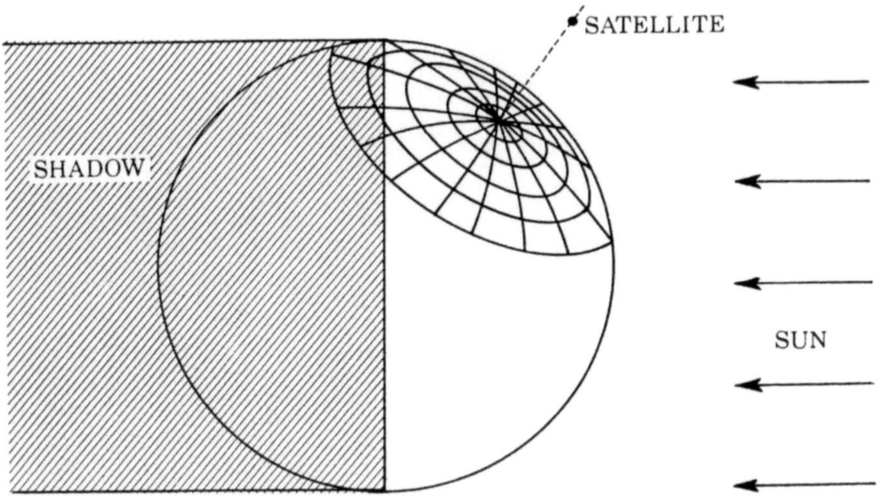

Figure 1. Division of the spherical cap of the subsatellite point in small parts.

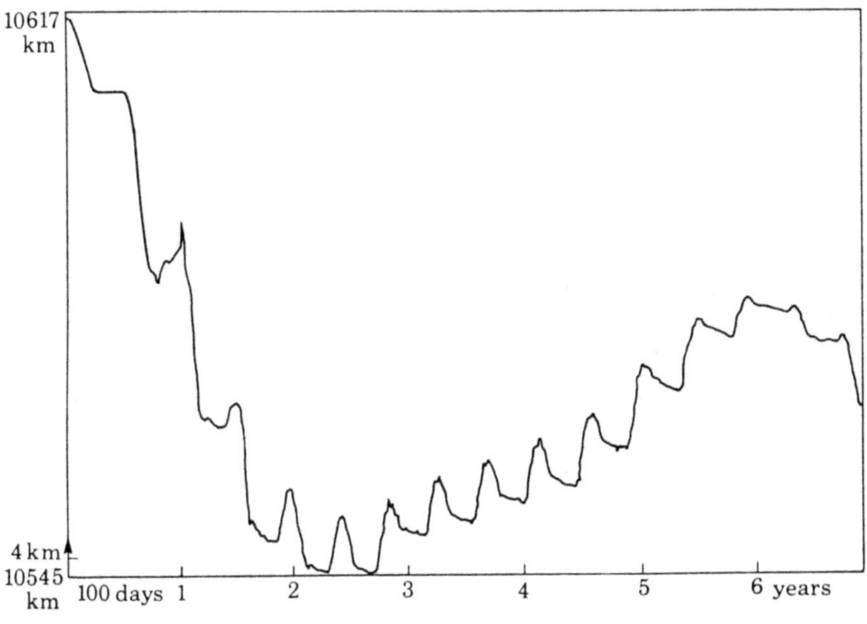

Figure 2.
Observed variations of the semi-major axis for seven years.

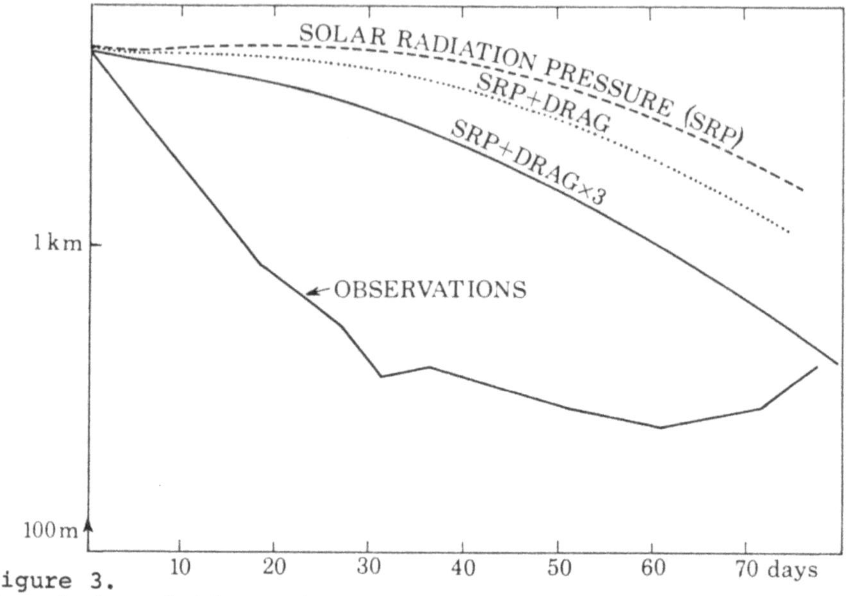

Figure 3.
Computed variations of the semi-major axis due to the direct solar pressure and to the air drag. These variations are compared with the observed variations.

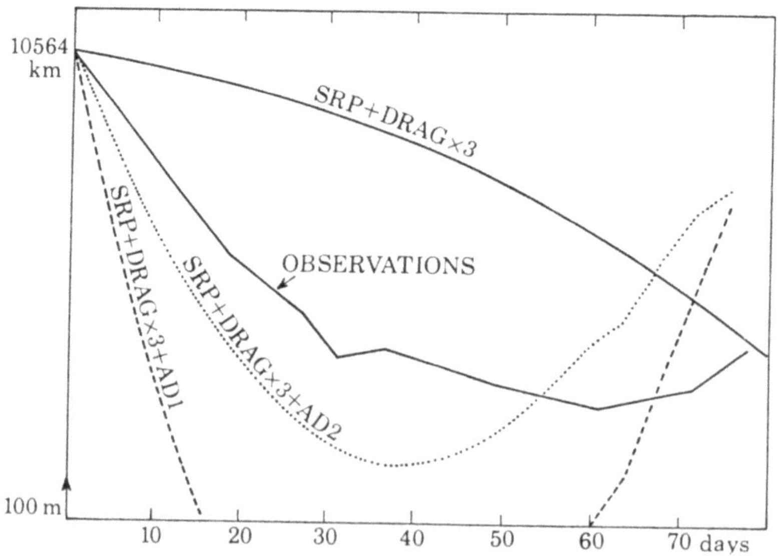

Figure 4.
Comparisons between observed variations of the semi-major axis and computed variations including the diffuse component (Wyatt model).

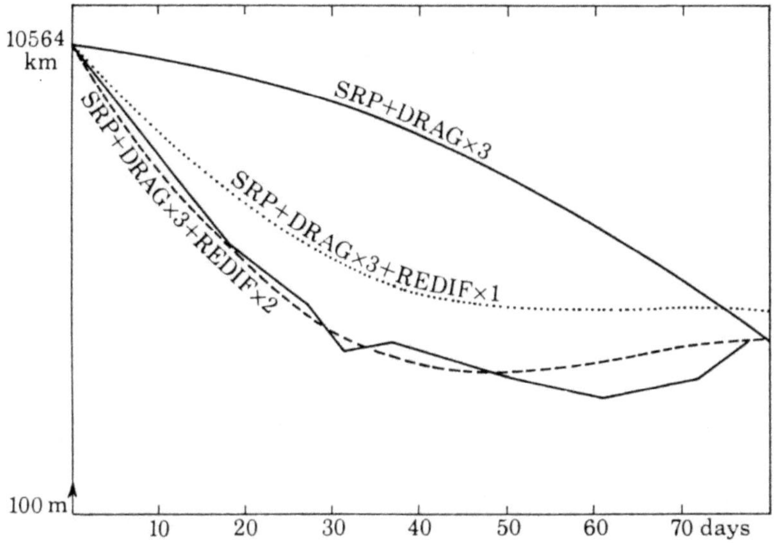

Figure 5.
Comparison between observed variations of the semi-major axis and computed variations including the diffuse component (Gambis model).

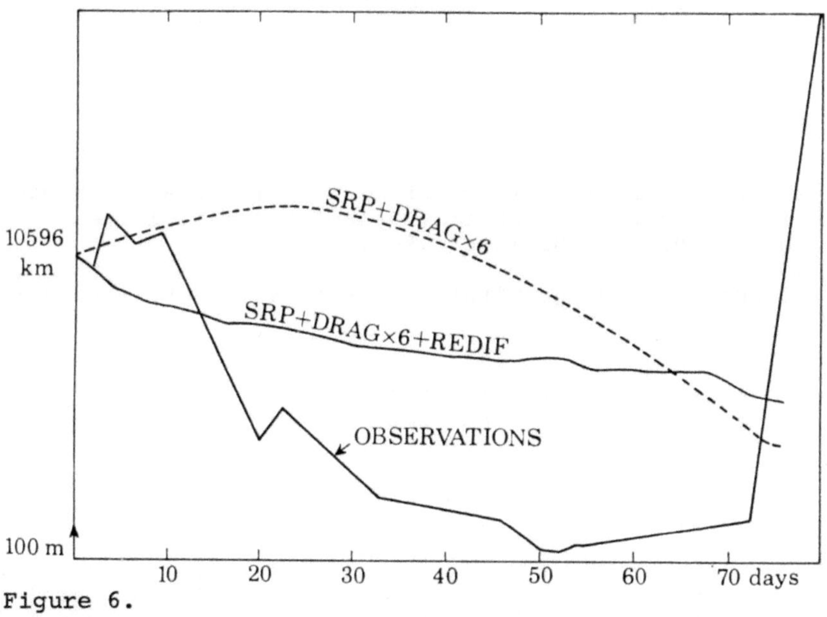

Figure 6.
An example where the agreement between observed and theoretical variations has not been observed.

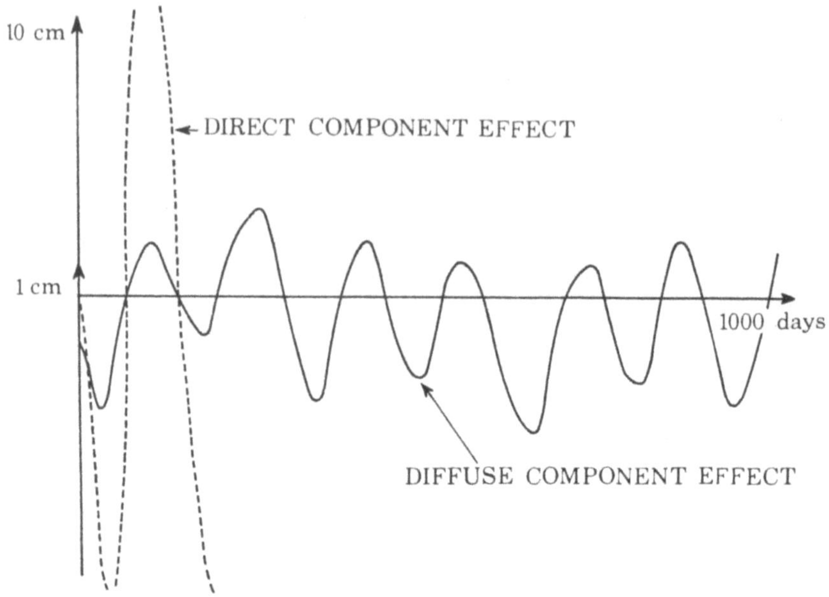

Figure 7.
Variations of the semi-major axis of the STARLETTE satellite due to the diffuse component. Comparison with direct solar radiation pressure.

References

1 Wyatt, S.P.: The effect of terrestrial radiation pressure on satellite orbits. Dynamics of satellites, ed. by M. Roy. Springer 1963.

2 Sehnal, L.: Nongravitational forces in satellite dynamics. This volume.

3 Rousseau, M.: Determination of densities at very high altitudes. Planet. and Space Sci. (1973).

4 Slowey, J.W.: Earth radiation pressure and the determination of density from atmospheric drag. Space Res. XIV, 1974 (in press).

5 Gambis, D.: Effet de la pression de radiation solaire rediffusée par la Terre sur le mouvement d'un satellite artificiel. Thèse de troisième cycle, Université de Paris VI (1974).

MIX
Papier aus verantwortungsvollen Quellen
Paper from responsible sources
FSC® C105338

If you have any concerns about our products,
you can contact us on
ProductSafety@springernature.com

In case Publisher is established outside the EU,
the EU authorized representative is:
Springer Nature Customer Service Center GmbH
Europaplatz 3, 69115 Heidelberg, Germany

Printed by Libri Plureos GmbH
in Hamburg, Germany